Understanding statistical error

A primer for biologists

Marek Gierliński
University of Dundee

WILEY Blackwell

This edition first published 2016 © 2016 by John Wiley & Sons Ltd

Registered office: John Wiley & Sons, Ltd, The Atrium, Southern Gate, Chichester,
West Sussex, PO19 8SQ, UK

111 River Street, Hoboken, NJ 07030-5774, USA

For details of our global editorial offices, for customer services and for information about how to
apply for permission to reuse the copyright material in this book please see our website at
www.wiley.com/wiley-blackwell.

Library of Congress Cataloging-in-Publication Data

Gierliński, Marek, author.
Understanding statistical error: a primer for biologists / Marek Gierliński.
    p. ; cm.
  Includes bibliographical references and index.
  ISBN 978-1-119-10691-3 (pbk.)
  I. Title.
[DNLM: 1. Statistics as Topic.   2. Analysis of Variance.   3. Biostatistics.   4. Computational
Biology–methods.   5. Probability.   6. Statistical Distributions. WA 950]
  R853.S7
  610.72'7–dc23

                                                                        2015024748

A catalogue record for this book is available from the British Library.

Wiley also publishes its books in a variety of electronic formats. Some content that appears in
print may not be available in electronic books.

Cover image: © Lonely_/iStockphoto

Set in 9/11pt PalatinoLtStd by Aptara Inc., New Delhi, India
Printed and bound in Malaysia by Vivar Printing Sdn Bhd

1   2016

Errors, like straws, upon the surface flow;
He who would search for pearls must dive below
—*John Dryden (1631–1700)*

Contents

Introduction 1
Why would you read an introduction? 1
What is this book about? 1
Who is this book for? 2
About maths 2
Acknowledgements 3

Chapter 1   Why do we need to evaluate errors? 4

Chapter 2   Probability distributions 7
2.1   Random variables 8
2.2   What is a probability distribution? 9
       Probability distribution of a discrete variable 9
       Probability distribution of a continuous variable 10
       Cumulative probability distribution 11
2.3   Mean, median, variance and standard deviation 11
2.4   Gaussian distribution 13
       Example: estimate an outlier 15
2.5   Central limit theorem 16
2.6   Log-normal distribution 18
2.7   Binomial distribution 20
2.8   Poisson distribution 23
       Classic example: horse kicks 25
       Inter-arrival times 26
2.9   Student's t-distribution 28
2.10   Exercises 30

Chapter 3   Measurement errors 32
3.1   Where do errors come from? 32
       Systematic errors 33
       Random errors 34
3.2   Simple model of random measurement errors 35
3.3   Intrinsic variability 38
3.4   Sampling error 39
       Sampling in time 39
3.5   Simple measurement errors 41
       Reading error 41
       Counting error 43
3.6   Exercises 46

Chapter 4   Statistical estimators 47
4.1   Population and sample 47
4.2   What is a statistical estimator? 49

4.3	Estimator bias	52
4.4	Commonly used statistical estimators	53
	Mean	53
	Weighted mean	54
	Geometric mean	55
	Median	56
	Standard deviation	57
	Unbiased estimator of standard deviation	59
	Mean deviation	62
	Pearson's correlation coefficient	63
	Proportion	65
4.5	Standard error	66
4.6	Standard error of the weighted mean	70
4.7	Error in the error	71
4.8	Degrees of freedom	72
4.9	Exercises	73

Chapter 5   Confidence intervals   **74**

5.1	Sampling distribution	75
5.2	Confidence interval: what does it really mean?	77
5.3	Why 95%?	79
5.4	Confidence interval of the mean	80
	Example	83
5.5	Standard error versus confidence interval	84
	How many standard errors are in a confidence interval?	84
	What is the confidence of the standard error?	85
5.6	Confidence interval of the median	86
	Simple approximation	89
	Example	89
5.7	Confidence interval of the correlation coefficient	90
	Significance of correlation	93
5.8	Confidence interval of a proportion	95
5.9	Confidence interval for count data	99
	Simple approximation	102
	Errors on count data are not integers	102
5.10	Bootstrapping	103
5.11	Replicates	105
	Sample size to find the mean	108
5.12	Exercises	109

Chapter 6   Error bars   **112**

6.1	Designing a good plot	112
	Elements of a good plot	113
	Lines in plots	115
	A digression on plot labels	116
	Logarithmic plots	117
6.2	Error bars in plots	118
	Various types of errors	119
	How to draw error bars	120

	Box plots	121
	Bar plots	123
	Pie charts	128
	Overlapping error bars	128
6.3	When can you get away without error bars?	130
	On a categorical variable	130
	When presenting raw data	130
	Large groups of data points	130
	When errors are small and negligible	131
	Where errors are not known	131
6.4	Quoting numbers and errors	132
	Significant figures	132
	Writing significant figures	133
	Errors and significant figures	135
	Error with no error	137
	Computer-generated numbers	138
	Summary	140
6.5	Exercises	140

Chapter 7	**Propagation of errors**	**142**
7.1	What is propagation of errors?	142
7.2	Single variable	143
	Scaling	144
	Logarithms	144
7.3	Multiple variables	146
	Sum or difference	146
	Ratio or product	147
7.4	Correlated variables	149
7.5	To use error propagation or not?	150
7.6	Example: distance between two dots	151
7.7	Derivation of the error propagation formula for one variable	153
7.8	Derivation of the error propagation formula for multiple variables	155
7.9	Exercises	157

Chapter 8	**Errors in simple linear regression**	**158**
8.1	Linear relation between two variables	158
	Mean response	159
	True response and noise	160
	Data linearization	161
8.2	Straight line fit	161
8.3	Confidence intervals of linear fit parameters	164
	Example	168
8.4	Linear fit prediction errors	170
8.5	Regression through the origin	173
	Example	174
8.6	General curve fitting	175
8.7	Derivation of errors on fit parameters	178
8.8	Exercises	179

Chapter 9   Worked example   **181**
9.1   The experiment   181
9.2   Results   182
        Sasha   183
        Lyosha   186
        Masha   189
9.3   Discussion   190
9.4   The final paragraph   192

Solutions to exercises   **193**
Appendix A   **206**
Bibliography   **209**
Index   **211**

Introduction

Why would you read an introduction?

It is common that each nonfiction book is preceded by an 'introduction', or a 'preface', or a 'foreword' or sometimes a combination of the above. If you are (un)lucky, you might find a note from the Editor, a foreword followed by the preface to the first edition, a preface to the second edition and a general introduction. There, first of all, you can read about how great the author is. Next, you will find that the book is unique and better than all other books on the topic written so far. Then, the author will delve into painstakingly detailed description of each chapter, which by the way can be found in the table of contents. Finally, there is time for compulsory acknowledgements to all family and friends who the author forced into reading his or her *magnum opus*. There is no escaping; forewords, prefaces and introductions are everywhere. Stanisław Lem once wrote a book consisting entirely of forewords (Lem 1979).

People usually skip all of these intros as they are boring, pretentious, self-righteous and useless. All right, are you still with me? If you managed to get that far, you might be one of the few who actually *read* introductions. Very well, then. I'll try to be brief, down to the point and not too conceited.

What is this book about?

As the title suggests, the book is about error analysis, with emphasis on applications in biology or, more generally, in life sciences. Since the time of the great Ronald Fisher, statistics have become an inherent part of biology. Very few numerical results from either biological or medical studies can make their way into publication without confirming their statistical significance. One way of doing this is by providing a p-value from a statistical test, or – roughly speaking – a probability of being wrong in a particular statement. That is what this book is *not* about.

Understanding Statistical Error: A Primer for Biologists, First Edition. Marek Gierliński.
© 2016 John Wiley & Sons, Ltd. Published 2016 by John Wiley & Sons, Ltd.

The other way of assessing the significance of a result is by finding its inherent error, or uncertainty. In my mind, a numerical result quoted without any kind of uncertainty is meaningless. Hence, it is good to know *how* to calculate errors. And that is what the book *is* about.

Here I discuss various aspects of error analysis: a bit of theoretical background and practical ways of calculating confidence intervals, but also graphical presentation of error bars and quoting numbers with errors. I put emphasis on intuition and understanding rather than practical computational recipes, although I give exact formulae for types of errors. Beware: this is not a comprehensive book on statistics; it is rather focused on practical understanding of uncertainty analysis. You can find more details in the table of contents, right after the introduction.

Who is this book for?

This book is written for an inquisitive biologist who wants to improve his or her understanding of data analysis. While a biologist is my target reader, the book may be useful for anyone who deals with numerical data and wants to learn more about how to evaluate and compare measurements. If you calculate various types of errors using a software package and you would like to find out where these errors come from, this book is for you. If you use standard deviations, standard errors and confidence intervals, but you are not sure what they really mean, this book is for you. If you struggle with finding errors of the median or correlation coefficient, this book is for you. Or, perhaps you are just curious and would like to learn a few basic things about uncertainty analysis – this book is also for you.

About maths

Despite the existence of a few attempts in the literature that use a purely intuitive approach (e.g. Motulsky 2010), I believe that it is very difficult to do statistics without maths. Plain English explanations cannot replace the strict precision of a mathematical equation. A simple derivation can explain where a given formula came from. Hence, there is maths in this book. Not very complex, not very extensive, but maths there is.

Needless to say, equations are required in practical applications, so if you need to find a particular uncertainty not provided by the statistical software you normally use, you can employ equations from this book. They can be easily encoded, either in any programming language or even in a computer spreadsheet. Mathematics in this book is quite basic; it doesn't really go beyond the level taught in a typical secondary school. Most equations contain simple algebra and sums. The most advanced operator I use is a derivative.

I don't want to scare potential readers away. This is not a mathematical textbook! I apply equations only when necessary and I always try to accompany them with an intuitive explanation. Often, I show the results of a computer simulation to illustrate the meaning of a concept or formula. I have also made a few simplifications and approximations here and there at the expense of mathematical correctness. I hope this makes the maths in this book much easier to understand.

I need to finish with a caveat. This is a book written primarily for biologists, not for mathematicians or physicists. Hence, there are no mathematical proofs, some derivations are not strict and there is a general lack of mathematical rigour. A mathematician might scowl at the content of this book, so if you are one, please shut your eyes now.

Acknowledgements

I would like to thank Professor Angus Lamond, who carefully read the manuscript from cover to cover and gave me a great deal of invaluable comments. Being a biologist, he helped me to understand better my target reader (you!). He also helped me with my English, which is not my first language.

Chapter 1

Why do we need to evaluate errors?

A measurement without error is meaningless.

—My physics teachers

Think of a number, a measurement from an experiment. We can determine in a microarray experiment, for example, levels of gene expression following a treatment of interest. Let us assume the resulting number is 19,086. It represents the intensity from a gene probe expressed in some arbitrary units. This number by itself doesn't tell us much. We need to compare it with a result from the control sample. Let's say the control gives an intensity of 39,361 for the same gene.

Looking at these two numbers, you might conclude that there is a twofold change in gene expression, and we all know that a twofold change is compelling. So, the gene of interest is suppressed under the treatment. Excellent! Time to publish the results.

But not so fast. The problem is that each measurement has an inherent uncertainty, or *error*. There is a limit as to how sure we can be that the experimental result is reflecting the true parameter we are trying to assess, in this case the level of gene expression. In some types of experiments, uncertainties can be high, so having two 'naked' numbers without knowing how robust they are doesn't mean the observed twofold change between our two conditions has any significance.

Now imagine you have a lot of money and a lot of time, and you can repeat your experiment (both control and treatment) 30 times. Each time, you measure expression of the same gene. The result is shown in Figure 1-1.

Understanding Statistical Error: A Primer for Biologists, First Edition. Marek Gierliński.
© 2016 John Wiley & Sons, Ltd. Published 2016 by John Wiley & Sons, Ltd.

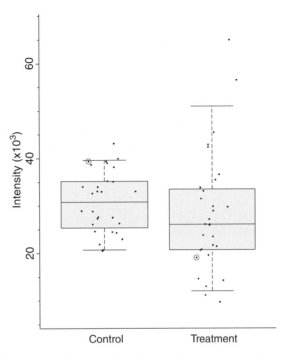

Figure 1-1. Control (left) and treatment (right) samples from an imaginary microarray experiment. Each measurement was done in 30 replicates. Clouds of points represent individual measurements; boxes encompass data between the 25th and 75th percentiles; whiskers span between the 5th and 95th percentiles. The line in the middle represents the sample median. Although the two initial measurements (circled points) differ by factor two, there is no statistically significant difference between the samples.

It turns out that repeated measurements of the same quantity reveal a huge scatter in the values obtained, with the results for control and treatment largely overlapping. This is not atypical in biology. You can aggregate your repeated results (a sample) and represent them by calculating the sample mean and standard error of the mean. These results are $(30.7 \pm 1.2) \times 10^3$ and $(28.3 \pm 2.3) \times 10^3$ for control and treatment, respectively. Now we have not only numbers, which come from repeated experiments, but also errors that represent the uncertainties of our measurements. These errors overlap, and a proper statistical test (e.g. a t-test) shows that there is no statistically significant difference between the mean value of the treatment and control ($p = 0.2$). The previous simplistic conclusion that the treatment changed the level of gene expression has, therefore, been shown to be incorrect.

A measurement without quoted error is meaningless.

This little example demonstrates why we need errors and error bars. In this book, I will explain how to evaluate errors the easy way. I will begin with basic concepts of probability distributions.

Chapter 2

Probability distributions

> Misunderstanding of probability may be the greatest of all impediments to scientific literacy.
>
> —*Stephen Jay Gould*

Consider an experiment in which we determine the number of viable bacteria in a sample. To do this, we can use a simple technique of dilution plating. The sample is diluted in five consecutive steps, and each time the concentration is reduced 10-fold. After the final step, we achieve the dilution of 10^{-5}. The diluted sample is then spread on a Petri dish and cultured in conditions appropriate for the bacteria. Each colony on the plate corresponds to one bacterium in the diluted sample. From this, we can estimate the number of bacteria in the original, undiluted sample.

Now, think of exactly the same experiment, repeated six times under the same conditions. Let us assume that in these six replicates, we found the following numbers of bacterial colonies: 5, 3, 3, 7, 3 and 9. What can we say about these results?

We notice that replicated experiments give different results. This is an obvious thing for an experimental biologist, but can we express it in more strict, mathematical terms? Well, we can interpret these counts as realizations of a *random variable*. But not just any *completely* random variable. This variable would follow a certain law, a Poisson law in this case. We can estimate and theoretically predict its *probability distribution*. We can use this knowledge to predict future results from similar experiments. We can also estimate the uncertainty, or error, of each result.

Firstly, I'm going to introduce the concept of a random variable and a probability distribution. These two are very closely related. Later in this chapter, I will show examples of a few important probability distributions, without which it would be difficult to understand error analysis.

Understanding Statistical Error: A Primer for Biologists, First Edition. Marek Gierliński.
© 2016 John Wiley & Sons, Ltd. Published 2016 by John Wiley & Sons, Ltd.

2.1 Random variables

I will not go into gory technical details. A random variable is a mathematical concept, and it has a formal definition. For the purpose of this book, let us say that a random variable can take random values. It sounds a bit tautological, but this is probably the simplest possible definition. In practice, a random variable is a result of an experiment. Its randomness manifests itself in the differing values of repeated measurements of the same quantity. It is quite common that each time you make your measurement, you obtain a different number.

> A random variable is a numerical outcome of an experiment. It will vary from trial to trial as the experiment is repeated.

Consider this example. Let us throw two dice and calculate the sum of the numbers shown. This can be any number between 2 and 12. More importantly, some results are more likely than others. For example, there is only one way of getting a 12 (a double 6), but there are five different combinations resulting in the sum of 6 ($1 + 5, 2 + 4, 3 + 3, 4 + 2$ and $5 + 1$). It is easy to see that throwing a 6 is five times more likely than throwing a 12.

An example of a non-random variable could be the number of mice used in an experiment. If you have five mice, you have five mice and the result stays unless you drink too much whisky and begin to see little white mice everywhere.

Hold on. In Chapter 1, I showed an example of a repeated measurement that gave a different value each time. So, what is going to happen if you repeat your murine experiment many times? Well, if you come back to the cage after a minute, you are quite likely to find five mice again (unless you forgot to lock the cage). The result is not going to change regardless of how many times you count them. This type of repeated measurement is called pseudo-replication.

More about replication and pseudoreplication in Section 5.11.

But this is not what we are asking about. Typically, you would be conducting an experiment (e.g. testing a drug), spanning over many days in which you would record mice dying and surviving. If you were to repeat the *entire* experiment many times, you might find that 10 days after dosing the mice with a particular drug there are three mice surviving in experiment 1, two mice alive in experiment 2, four in experiment 3 and so on. Although your particular measurement (counting mice) is 'perfect' and not biased by any error, the repeated experiments show the actual level of uncertainty. Hence, contrary to simple intuition, the number of mice at any given moment of time is a random variable. Most values in biological experiments are random variables.

There are two kinds of random variables: discrete and continuous. *Discrete* random variables can take only certain values, typically whole numbers. The number of mice is a discrete variable, as it can only be 0, 1, 2, 3 and so on. Alternatively, discrete values might be categorical, for example male/female. If necessary, categories can be converted into integer numbers. In contrast, *continuous* random variables can take any values, typically any real numbers. The length of a mouse's tail is an example of a continuous variable.

2.2 What is a probability distribution?

Every random variable obeys a specific statistical law, called a probability distribution. As the name suggests, this law tells us how the random variable is distributed. Or, to convey it more precisely,

> A probability distribution defines the probability of finding the random variable within a certain range of values.

I will use the following notation in this section. A random variable (X) is denoted by a capital letter. This is only a name. Small letters (k, x) denote possible values that the random variable can take. These are actual numbers.

Probability distribution of a discrete variable

Let us consider a discrete random variable X, which can assume non-negative integer values 0, 1, 2, 3, ... I will denote $P(X = k)$ as the probability of the variable X being equal to the value k. Mathematically speaking, the probability of finding X between two numbers a and b is determined by the following equation:

$$P(a \leq X \leq b) = \sum_{k=a}^{b} P(X = k), \qquad (2\text{-}1)$$

which is, simply, the sum of all individual probabilities. For example, in Figure 2-1a, three shaded bars show probabilities of $P(X = 5) = 0.16$, $P(X = 6) = 0.10$ and $P(X = 7) = 0.06$. The sum

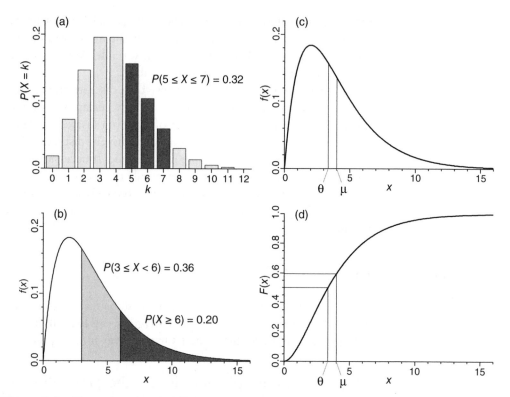

Figure 2-1. Examples of probability distributions. (a) Distribution of a discrete random variable X, where each bar shows the probability of X being equal to k. (b) Continuous distribution, probability of finding X between two values equals the area under the f(x) curve between these two values. (c) The same distribution as in (b), with median, θ, and mean, μ, marked. (d) Cumulative distribution, F(x), corresponding to the distribution f(x) from panel (c). By definition, $F(\theta) = 0.5$.

of these probabilities is 0.32. Hence, $P(5 \leq X \leq 7) = 0.32$. The total probability over all possible values of X is always unity: $P(0 \leq X \leq \infty) = 1$.

Probability distribution of a continuous variable

A continuous random variable X can take on any real value x. Here we use a probability *density* function, $f(x)$, which defines the probability per unit x. As such, the value of this function for any specific x doesn't have a simple intuitive meaning. It only makes sense when integrated (or summed up) over a certain range:

$$P(a \leq X \leq b) = \int_a^b f(x)dx, \tag{2-2}$$

Graphically, this integral corresponds to an area under the curve $f(x)$ between a and b, as shown in Figure 2-1b. The probability of finding X between 3 and 6 is indicated by the light-shaded area and equals $P(3 \leq X < 6) = 0.36$. The dark-shaded region shows the probability of X being greater (or equal to) 6, $P(X \geq 6) = 0.20$. The interval is from 6 to infinity. The total probability over all possible values of X is always unity: $P(-\infty \leq X \leq \infty) = 1$.

If we narrow the range of integration to nothing ($a = b$), the resulting probability is zero, as the area under the curve collapses to nothing. Hence, $P(X = 5) = 0$ in a continuous distribution. Because X is as a continuous variable, it can assume an infinite number of values in any arbitrary interval around 5, so the chances of hitting exactly 5 (I mean *exactly*) is infinitesimally small.

Cumulative probability distribution

Another useful function is a *cumulative probability distribution*, defined as the probability that some random variable X is less than x: $F(x) = P(X < x)$. It can be graphically represented as the area under the curve f to the left of x. Due to this definition, $F(x)$ is a monotonic[1] function, growing from 0 to 1, with a characteristic 'sigmoid' shape in the plot. An example of a probability density function, $f(x)$, and its cumulative distribution, $F(x)$, is shown in Figure 2-1c and 2-1d. It can be understood as a left-tail probability, that is, $P(X < x)$. The right-tail probability is then $P(X \geq x) = 1 - F(x)$. These two terms are often used in statistical tests and for finding confidence intervals. Many probability distributions (e.g. the Student's t-distribution in the Appendix) are tabulated as cumulative distributions, typically as right-tail probabilities.

Cumulative distribution is a left-tail probability.

2.3   Mean, median, variance and standard deviation

A probability distribution of the random variable X tells us everything we want to know about this variable. Sometimes, however, we would like to reduce this knowledge to a simple number, that

---

[1]A function $F(x)$ is monotonic if it grows with growing x. A curve of F plotted versus x always goes up (or stays level) with increasing x.

is, a *parameter* describing one particular aspect of the distribution. For example, we might want to know where it is centred, or how wide it is. In this section, I show a few such parameters: mean, median, variance and standard deviation.

Population, sample and statistical estimators are discussed in Chapter 4.

Please note that these are quantities calculated for a theoretical probability distribution. In practice, we often use them to describe *population* properties, assuming that the population is, for example, Gaussian. In contrast, mean, median, standard deviation (and so on) calculated for a *sample* are called *statistical estimators.* Sample parameters are found using similar, but not necessarily identical, equations and should not be confused with the equations given in this section. Sample parameters only approximate population parameters. I will explain this later in the book.

For the sample mean, see Section 4.4, subsection 'Mean'.

For each random variable X, we can define the *mean* (also called the *expected value*) of X as

$$\mu = \sum_{k=0}^{\infty} kP(X = k), \tag{2-3}$$

or

$$\mu = \int_{-\infty}^{\infty} xf(x)dx, \tag{2-4}$$

for a discrete and continuous variable, respectively. The mean of a random variable X is often denoted as $\overline{X}$ or $\langle X \rangle$. Equations (2-3) and (2-4) can be interpreted as *weighted* means, where weights are either individual probabilities in the case of a discrete variable, or the probability density for a continuous variable.

For the sample median, see Section 4.4, subsection 'Median'.

The *median* of a random variable X is defined as a value θ, where $F(\theta) = 0.5$. It divides the distribution into two halves with equal probability, $P(X \le \theta) = P(X \ge \theta) = 0.5$. The vertical lines in Figure 2-1c show the mean and the median. These two quantities are equal for a symmetric distribution, such as a Gaussian distribution, as discussed in Section 2.4.

For the sample standard deviation see Section 4.4, subsection 'Standard deviation'.

Having found the mean, we can define a *residual* as a displacement from the mean, $X - \mu$. *Variance* is then defined as the mean squared residual,

$$\sigma^2 = \langle (X - \mu)^2 \rangle. \tag{2-5}$$

The square root of the variance, $\sigma = \sqrt{\sigma^2}$, is called the *standard deviation* and is one of the most important quantities in error analysis. It measures dispersion of X and typically describes the uncertainty of a measurement.

2.4 Gaussian distribution

The Gaussian distribution is commonly observed in nature, and because of that most people call it *normal*. For some reason physicists call it *Gaussian*, even though it was a French mathematician, Abraham de Moivre (1667–1754), who published its mathematical formulation many years before Gauss (1777–1855) was even born. I'm a physicist by nurture, so I'm going to call it Gaussian most of the time, although I might occasionally confuse the reader by using the other term.

A graph of the Gaussian distribution is often termed a *bell curve* and is described by the following function:

$$f(x) = \frac{1}{\sqrt{2\pi\sigma^2}} e^{-\frac{1}{2}\left(\frac{x-\mu}{\sigma}\right)^2}, \tag{2-6}$$

where μ is the mean and σ is the standard deviation. These quantities are as defined in Section 2.3: if you take $f(x)$ from equation (2-6) and plug it into equation (2-4), you will actually get μ. If you calculate the mean squared residual, by replacing x with $(x - \mu)^2$ in equation (2-4), you will get the variance, σ^2. The quantity $Z = (x - \mu)/\sigma$ is often called a Z-score, and it represents the distance of x from the mean in units of standard deviation. The Gaussian curve is symmetric with respect to the mean, and its width is characterized by σ.

An example of the Gaussian function is shown in Figure 2-2. With increasing and decreasing mean, the bell-shaped curve shifts either to the right or to the left; with increasing and decreasing standard deviation, the curve becomes either broader or narrower. I will denote the Gaussian (normal) distribution of mean μ and standard deviation σ as $\mathcal{N}(\mu, \sigma)^2$.

Gaussian distribution of random errors is shown in Section 3.2.

Many quantities observed and measured in biology obey the Gaussian distribution. For example, the distribution of heights of a sufficiently large population of people is Gaussian. More importantly for us, many errors are normally distributed (i.e. show a Gaussian distribution; more about this later). Hence, in order to find some confidence limits, we need to know *probabilities* either within, or outside, certain intervals.

Gaussian probabilities are summarized in Table 2-1. Probabilities of being within one, two and three sigma are also shown in

---

[2]A typical notation used in literature would contain the mean and the variance, $\mathcal{N}(\mu, \sigma^2)$, but since I use the standard deviation a lot in this book, I decided to choose a nonstandard notation, $\mathcal{N}(\mu, \sigma)$.

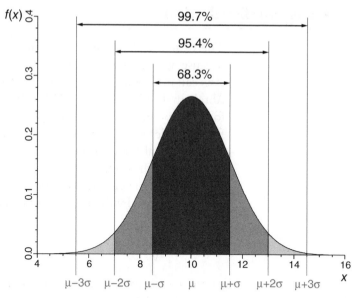

Figure 2-2. Gaussian distribution with mean, $\mu = 10$ and standard deviation, $\sigma = 1.5$. The arrows and numbers above the curve show one, two and three sigma ranges and the corresponding probabilities.

Figure 2-2. For example, there is a ~68% probability of finding a random Gaussian variable within $\mu \pm 1\sigma$. This is why roughly two-thirds of normally distributed data points are within one standard deviation from the mean. But it works only when the distribution approximates Gaussian (see Section 2.6). Obviously, the probability outside this range (i.e. below $\mu - \sigma$ or above $\mu + \sigma$) is ~32%.

Table 2-1. Probabilities for the Gaussian distribution. The first column shows the range with respect to the mean; the second and third columns show probabilities of finding a Gaussian random variable inside and outside this range, respectively. The last column shows approximate odds of being outside the range.

Range	In	Out	Odds
$\pm 1\sigma$	68.3%	31.7%	1:3
$\pm 1.96\sigma$	95.0%	5.0%	1:20
$\pm 2\sigma$	95.4%	4.6%	1:20
$\pm 2.58\sigma$	99.0%	1.0%	1:100
$\pm 3\sigma$	99.7%	0.3%	1:400
$\pm 4\sigma$	99.994%	0.006%	1:16,000
$\pm 5\sigma$	99.99994%	0.00006%	1:1,700,000

The meaning of the 95% confidence interval (95% CI) is explained in Section 5.2.

A 95% confidence interval (95% CI) which equates to a 5% p-value limit is commonly used in biology. From Table 2-1 you can see that this corresponds to the interval $\mu \pm 1.96\sigma$. This number, 1.96, is worth remembering. It roughly corresponds to a 'two sigma' result. Many physicists wouldn't call a result significant unless it is better than three sigma. Hence you can often hear about *three sigma significance*. It corresponds to the 'out' probability (or a p-value) of 0.3%. This also highlights a difference between the biology and physics communities. Biologists are willing to accept a result which might be wrong on average once in 20 cases. Physicists usually pull this limit down to 1 in 400. Whichever value you prefer, what is really important is that you understand the assumptions you are making and their limitations.

Example: estimate an outlier

Consider a study on obesity in which we weigh 100 mice. The mean weight from this sample is 20 g and the standard deviation is 5 g. One particular mouse of interest, let's call it Jerry, weighs 30 g. We want to know how Jerry stands against the entire population, or, more precisely, what is the probability (p-value) of obtaining such embarrassingly high body mass purely by chance, and not as a result of either his genetic constitution or the environment he grew in. Or, to express it in terminology used in biology: is Jerry an outlier?

It seems very simple. Jerry is two standard deviations away from the mean weight. From Table 2-1 we find that the probability of being further than 2σ away from the mean is 4.6%. Hence, this is our p-value. But is it exactly what we are looking for? We can ask two questions here. Is Jerry's body mass significantly *different* from that of a typical mouse? Or, is his mass significantly *higher* than that of a typical mouse? In the former case, we are looking for a two-tail probability (light-shaded area on both sides in Figure 2-2) and the resulting p-value is 4.6%. In the latter case, we are only interested in the right-hand tail, and because the Gaussian distribution is symmetric, the resulting p-value is 2.3%. Is it significant?

Well, it depends. The result tells us that Jerry is (roughly) in the top 2% of the population when it comes to body weight. Clearly, he is a fatso. On the other hand, in a sample of 100 mice, you'd expect on average about two little monsters weighing more than 30 g, just because the probability of this event is about 2%. In high-throughput biological experiments, it is typical to have hundreds of thousands or even millions of measurements. You should not be surprised to come across some individual values standing out four or five standard deviations from the mean.

> Rare events are expected to be seen in large samples.

In this context, if you ask whether Jerry's weight is unexpected in the sample, the answer is no. Although he is hopelessly fat, he doesn't have to be viewed as an outlier in a statistical sense.

> In statistics, an outlier is defined as an observation that stands out from the sample.

There are methods, or rather rules of thumb, to decide if the given observation is an outlier or not. *Chauvenet's criterion* states that if the Gaussian tail probability for the given point multiplied by the sample size is less than 0.5, the point in question is an outlier. In our case we multiply 0.023 by 100 and get 2.3 > 0.5. According to Chauvenet's criterion, Jerry is not an outlier[3]. The controversial bit of this method is that it recommends removing outliers from the sample before further processing. In biology, we are usually interested in outliers, because they show interesting behaviour. Therefore, if you decide to use Chauvenet's criterion to 'clean up' your data, you had better know what you are doing.

2.5   Central limit theorem

This is one of the most important theorems in statistics. It has significant implications for error theory. Roughly speaking, it states that *the sum of a large number of any independent random variables is Gaussian*. And when I say *any*, I mean it: these individual variables can have any arbitrary probability distribution. When you add them up, original distributions are erased, and their sum is distributed normally. The same applies to the mean of all variables.

Let us have a look at the following example. Imagine that we throw N dice together and calculate the mean value of all the numbers shown, M_N. We want to know how M_N is distributed. The most obvious way of finding a distribution is to repeat the

---

[3]Mind you, the Gaussian probability drops very quickly as we go away from the mean. A 33-g mouse would be an outlier according to Chauvenet's criterion.

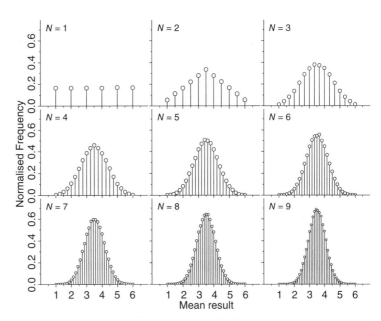

Figure 2-3. Results of throwing N dice together ($N = 1, 2, \ldots, 9$). Each panel shows the distribution of the mean value shown by N dice. Each distribution is calculated by throwing the dice 100,000 times. The observed frequencies (vertical axis) are normalized to 1.

experiment many times and plot a histogram of all the numbers obtained. For the purpose of this experiment, I threw the dice 100,000 times. Obviously, I didn't spend all my afternoon tossing little cubes; I did a computer simulation instead. The result is the same.

We start with $N = 1$. If the die is honest, the probability of obtaining each number between 1 and 6 is equal. So, after throwing it many times we should get a uniform distribution of numbers, as shown in Figure 2-3 (top left panel). The distribution is flat.

I talked about two dice as a random variable in Section 2.1. Now, let's try two dice and find the mean (the sum divided by 2). I have already mentioned this case before: getting the sum of 6 (mean = 3) is five times more likely than getting a 12 (mean = 6). This is illustrated in the second panel in Figure 2-3, where distribution of the mean has a triangular shape, with a peak in the middle. If we continue this experiment with ever increasing N, the shape of the distribution of the mean begins to resemble the bell curve of the Gaussian distribution.

Central limit theorem: the sum (or mean) of several random variables with arbitrary distributions approximates the Gaussian distribution.

I will discuss sampling distribution of the mean in Section 5.4. It might seem at the moment that the central limit theorem is only a pointless exercise in throwing dice, but I cannot overemphasize its importance for statistics and error analysis. It will become clearer when I discuss the sampling distribution of the mean and the distribution of random errors.

Another practical consequence of the central limit theorem is that certain distributions (some of them discussed later in this book) become approximately Gaussian for a large sample size. This is true for binomial, Poisson, χ^2 and Student's t-distributions.

2.6 Log-normal distribution

Log-normal distribution is a probability distribution of a random variable whose logarithm is normally distributed. In other words, if X is log-normally distributed, then $Y = \log X$ is normally distributed. And vice-versa: if Y is normally distributed, then $X = 10^Y$ is log-normally distributed. I'm sure this formal definition is not clear at all, so let us look at a real-life example.

Consider a proteomics experiment in which we find peptide ion intensities (a measure of peptide abundances) using a mass spectrometer. We have 93,338 data points and find the mean, $M = 2.1 \times 10^6$ (in arbitrary units). and the standard deviation, $SD = 7.4 \times 10^6$. Right away, we can notice that something is not quite right here: the standard deviation is huge, over three times larger than the mean. The best thing to do in such cases is to plot the distribution of your measurements. It is shown in Figure 2-4a.

You can immediately see that the distribution of peptide intensities is very asymmetric. It looks, however, very different if we take a logarithm (in this case using base 10, but that doesn't matter) of intensities and plot their distribution (Figure 2-4b). This vaguely resembles a Gaussian distribution, and its mean and standard deviation are $M_{\log} = 5.7$ and $SD_{\log} = 0.7$, respectively. We often refer to these two plots as having linear and logarithmic scale, and data as being in linear and logarithmic space, respectively.

Log-normal distribution is not uncommon in biological experiments. You can expect peptide or protein abundances estimated from a mass spectrometer to be log-normally distributed. Similar distributions are seen in gene expression experiments. Also, drug potency, IC_{50}, is typically log-normally distributed. If you encounter this kind of data, it is wise to take a logarithm of all data points before further processing.

One thing I need to mention when discussing distributions that are very far from Gaussian is how standard deviation can be used.

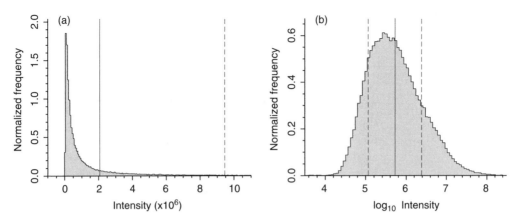

Figure 2-4. An example of log-normal distribution. These are peptide intensities from a mass-spectrometry experiment. (a) Normalized (to the total number of data points) distribution of intensities. (b) Normalized distribution of the logarithm of the same intensities. Solid vertical line shows the distribution mean, while dashed vertical lines mean ± standard deviation. The mean minus standard deviation is outside the figure in panel (a).

It seems to be common knowledge that 'you should expect about two-thirds of the data points to be within one standard deviation of the mean'. This is a true statement, but only under the assumption that data are normally distributed. From Table 2-1, we can see that the probability of a random variable being within one sigma of the mean is about 68%, roughly two-thirds. Hence, if your data follow the Gaussian distribution, you can expect the above statement to be true. This doesn't work for non-Gaussian distributions. For example, among our peptides, 96% of data points are within one standard deviation from the mean (i.e. within $M \pm SD$). This is much more than two-thirds. On the other hand, if we do the same calculation for the logarithms of intensities, it turns out that 67% of them are within $M_{\log} \pm SD_{\log}$. One should be very careful with some popular statements in statistics: they usually depend on certain assumptions that might or might not be true.

About two-thirds of data points are within one standard deviation from the mean *only* when their distribution is approximately Gaussian.

See a very brief discussion of the geometric mean in Section 4.4.

A few more comments regarding logarithms might be worth adding here. Firstly, it sounds quite obvious, but I need to mention that the mean of logarithms does not equal the logarithm of the mean. Our log-normal distribution example shows it clearly. The logarithm of the mean is $\log M = 6.3$, whereas the mean of logarithms is $M_{\log} = 5.7$. One cannot be replaced with the other.

Another note is on the logarithmic base in plots. Base 2 is very common in biology, probably because it shows the ratio of 2 in a natural way, as $\log_2 2 = 1$. However, there is nothing magical about the ratio of 2; it does not have any special statistical or biological meaning. As I demonstrated in Chapter 1, the ratio of 2 does not automatically make things significant. Using base 10, on the other hand, has an advantage in plots with logarithmic axes. When you see a 6 on the axis, you know it corresponds to a million (10^6). This is not the case in base 2 log plots; when you see a 12 on the axis, it is not immediately obvious that it represents 4096 in the linear space. Generally, it does not matter what logarithmic base you use, as all of them are equally valid. The important thing is to be consistent.

2.7 Binomial distribution

Let us begin with an example. Consider throwing a symmetric coin five times. What is the probability of obtaining heads exactly three times? When you toss a coin once, the probability of getting heads is 0.5. If you toss the coin again, the probability of having heads is 0.5 again. The result of the first throw does not affect the outcome of the second throw (they are independent). However, if you consider these two events *together* and ask about the probability of having heads *and* heads, the second throw gives a 0.5 probability out of the 0.5 from the first throw. Hence, the probability of having heads after heads is half out of a half, $0.5 \times 0.5 = 0.5^2 = 0.25$. If you add a third throw, the probability of having three heads would again be half of the probability of having two heads, $0.5^3 = 0.125$. This is a general rule in probability theory; for independent events A and B, the probability that *both* events occur is $P(A \text{ and } B) = P(A)P(B)$.

The probability of observing three heads is 0.5^3, but then we require that the remaining two throws result in tails. Following a similar logic as above, we deduce that the probability of having two tails is 0.5^2. Eventually, the probability of having heads three times *and* tails two times is the product of these two numbers, $0.5^3 \times 0.5^2 = 0.125 \times 0.25 = 0.03125$.

This seems to answer our question, but there is one more thing to consider. The combination of three heads and two tails can be obtained in more than one way, as illustrated in Figure 2-5,

Figure 2-5. Ten different ways of having three successes among five events. Each black circle indicates a success, while each white circle shows a failure. The number of these possibilities is described by the binomial coefficient, $\binom{5}{2} = \frac{5!}{3!(5-3)!} = \frac{5!}{3! \times 2!} = \frac{2 \times 3 \times 4 \times 5}{2 \times 3 \times 2} = \frac{4 \times 5}{2} = 10.$

where black circles represent heads and white circles represent tails. There are 10 ways of getting three heads and two tails, each of them equally likely with a probability of 0.03125. They are mutually exclusive (i.e. we can observe only one combination at a time), and either of these 10 combinations fulfils our requirement of observing exactly three heads. For two mutually exclusive events A and B, the probability that *either* of the events occurs is $P(A \text{ or } B) = P(A) + P(B)$. Hence, we need to add all these 10 individual probabilities. Finally, the probability of getting heads three times in a series of five coin throws is $10 \times 0.03125 = 0.3125$.

I will demonstrate applications of the binomial distribution in Sections 3.2 and 5.6.

The random variable representing the number of observed heads is described by the *binomial distribution*. More generally, we talk about the probability of a number of successes in a sequence of independent events. Each event (a flip of a coin in our example) can end up in either a 'success' or a 'failure'. If the probability of success in each event is p (in the coin example, $p = 0.5$, but it can be generalized to any probability), then the probability of having exactly k successes in n events is described by the following formula:

$$P(S = k) = \binom{n}{k} p^k (1 - p)^{n-k}. \tag{2-7}$$

S is a binomially distributed random variable. I will show applications of the binomial distribution later in the book.

But now, let me try to explain the intuition behind this equation. The first part,

$$\binom{n}{k} = \frac{n!}{k!(n-k)!}, \tag{2-8}$$

is called the *binomial coefficient*. The exclamation mark denotes a factorial, for example, $5! = 1 \times 2 \times 3 \times 4 \times 5 = 120$. It is often read 'n choose k', and describes the number of possible choices of k elements from a set of n elements, or the number of ways k successes can be distributed among n events. An example is shown in Figure 2-5.

Furthermore, in equation (2-7), p is the probability of success in a single event. As mentioned before, the probability that k successes occur is p^k. Since we want k successes, we must have $n - k$ failures. The probability of a single failure is $1 - p$; therefore, the probability of $n - k$ failures is $(1 - p)^{n-k}$. Taking these two things into account, we have the total probability of $p^k(1 - p)^{n-k}$. However, such combination (k successes and $n - k$ failures) can occur in many ways. There are $\binom{n}{k}$ ways of selecting k successes out of n events, so we have to multiply the total probability by this number, and in this way we get equation (2-7).

A random variable with binomial distribution has the mean

$$\mu = np, \tag{2-9}$$

and the standard deviation

$$\sigma = \sqrt{np(1 - p)}. \tag{2-10}$$

When $p = 0.5$ (e.g. flipping a coin), the distribution is symmetric. An example of such a distribution is shown in Figure 2-6. When p is far from 0.5, the distribution is skewed. However, when n is large enough, it becomes symmetric again and it approximates a Gaussian distribution.

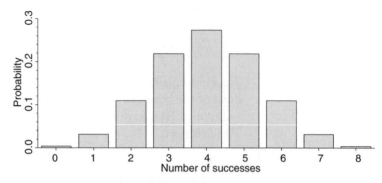

Figure 2-6. Binomial probability distribution showing probability of having k successes (horizontal axis) out of eight trials. The probability of success in a single trial is 0.5.

2.8 Poisson distribution

Consider radioactive decay. We know from physics that an atom can decay spontaneously and emit an ionizing particle. This process is stochastic, which means there is no way of predicting *when* this is going to happen. We can only predict *how likely* it is for an atom to decay over a certain period of time. From this, we can infer statistical properties of a group of atoms in terms of probabilities and expected decay rates. They will obey the Poisson distribution.

Imagine an experiment in which we register ionizing radiation from radionuclides undergoing radioactive decay. With a sensitive detector, we can count individual decay events. Let us do it for a while. Recorded times of decay events are marked with black dots in Figure 2-7a. Now we can group these events together and count the number of decays in one-second intervals (or bins). This

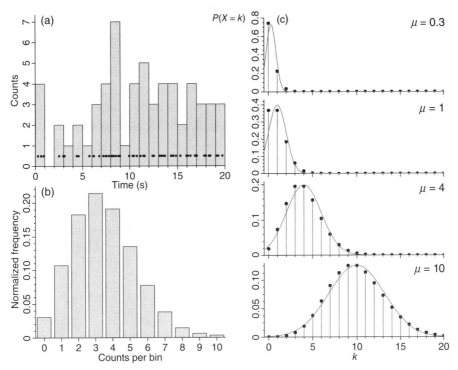

Figure 2-7. Poisson distribution. (a) Radioactive decay. Black dots show recorded times of individual decay events. These are counted in 1-s intervals (bins), and the count rate per bin is shown as a bar plot. The mean number of counts per bin is 3.5. (b) Distribution of the number of counts per bin, calculated over the long period of time. This approximates Poisson distribution with $\mu = 3.5$ counts per bin. (c) Examples of Poisson distribution for the mean rate of 0.3, 1, 4 and 10. Curves show Gaussian distribution with the same mean and $\sigma = \sqrt{\mu}$.

binning is shown as a bar plot in Figure 2-7a. As you can see, the number of counts per bin (or *count rate*) varies significantly from bin to bin. There are 7 counts in the 8–9 s bin, but there are no counts at all in the 1–2 s interval. This demonstrates the stochastic nature of radioactive decay.

The Poisson distribution is a distribution of counts.

Although we cannot predict how many counts will appear in the next second, we can find out how likely it is to observe a given number of counts. Let us use the results of this experiment (but carried out over a much longer time than the 20 s shown in Figure 2-7a) to build a distribution of the number of counts per bin. This is done simply by counting the bins with no decays, bins with one decay, with two decays and so on. The resulting frequency distribution (normalized, so the sum of all frequencies is 1) is shown in Figure 2-7b. Unlike individual bin count rates, this distribution is not random, and with a sufficiently large number of counts it always looks the same. It tells us that it is most likely to have three counts per bin, and it is rather unlikely to observe 10 counts over one second.

This is the Poisson distribution, and it is characterized by the mean number of counts per bin, μ. For the given mean rate, μ, the probability of observing exactly k counts in a bin is

$$P(X = k) = \frac{\mu^k e^{-\mu}}{k!}. \tag{2-11}$$

This function for a particular value of $\mu = 3.5$ counts per second is shown in Figure 2-7b. The probability of observing exactly three counts in a second is about 0.22 in this example. Note that the probability of finding no counts at all in a bin is non-zero and equals

$$P(X = 0) = \frac{\mu^0 e^{-\mu}}{0!} = \frac{1 \times e^{-\mu}}{1} = e^{-\mu}. \tag{2-12}$$

By definition[4], $0! = 1$. For $\mu = 3.5$, we find $P(X = 0) \approx 0.03$.

---

[4]You cannot multiply all the numbers from 1 up to zero, so mathematicians decided to define $0! = 1$. One of the reasons for this is for the binomial coefficient [equation (2-8)] to work correctly. There is only one way of choosing two successes out of two events, so we need $\binom{2}{2} = 1$. Hence, $\binom{2}{2} = \frac{2!}{2! \times 0!} = \frac{1}{0!} = 1$, and therefore $0! = 1$.

Unlike the Gaussian distribution, which is characterized by a mean and a standard deviation, there is only one free parameter here, μ. The standard deviation, which tells us how broad the distribution is, is the square root of the mean:

$$\sigma = \sqrt{\mu}. \tag{2-13}$$

See Section 3.5 for counting errors.

This is an interesting property and can be used to estimate errors or confidence intervals on a quantity that follows the Poisson distribution.

The Poisson distribution is not limited to events occurring in defined time intervals. It also can be applied to counts in a volume or space or a combination of space and time units. For example, the number of cell colonies growing on a Petri dish can be Poissonian. You would measure counts per area in such an experiment. Generally, a stochastic process is Poissonian if:

- events occur randomly,
- they are independent of each other, and
- the mean rate of events doesn't change over time or space.

Independence of events is particularly important. For example, motile cells can aggregate in clumps and their counts will not conform to the Poisson distribution. Suppose you count such clumped cells in different squares of a counting chamber. Sometimes you get a high result, and sometimes a very low one. The standard deviation (dispersion) of your counts is going to be very large, much larger than the square root of the mean. If you observe such behaviour, the distribution of the counts most likely is not Poissonian.

Another interesting property of the Poisson distribution is that for large count rates, it approximates a Gaussian distribution. The exact meaning of *large* is a bit vague, but I would say that in practical applications, 30 or so counts per bin would be large enough to consider the observed distribution to be normal. Figure 2-7c compares the shape of the Poisson and Gaussian distributions for the increasing value of the mean.

Classic example: horse kicks

This example comes from Ladislaus von Bortkiewicz (1868–1931), a Russian economist and statistician of Polish origin who lived and

Table 2-2.   Distribution of horse-kick deaths in the Prussian cavalry (Von Bortkiewicz 1898). The first column shows the number of horse-kick deaths in a 'bin' of one cavalry corps in one year. The second column shows the recorded number of such events during the entire study (20 years, 14 corps). The third column shows the predicted number of events from the Poisson distribution with the mean of 0.70 deaths per corps-year.

Deaths per corps-year	Frequency	Poisson prediction
0	144	139.0
1	91	97.3
2	32	34.1
3	11	7.9
4	2	1.4
5 or more	0	0.2

worked in Germany. In his book *Das Gesetz der kleinen Zahlen*[5], he analysed the number of soldiers in the Prussian cavalry killed by horse kicks. His analysis covered 14 cavalry corps over 20 years. He divided this into 280 individual corps-years (*bins*) and counted the frequency of horse-kick deaths in each bin (Table 2-2). This is analogous to the count rate in radioactive decay, although here frequencies are calculated in space–time bins. For example, there were no horse-kick deaths at all (count rate of zero) in 144 corps-years, and the unfortunate four deaths in one army corps in one year happened twice. The mean count rate is $\mu = 0.70$ deaths per corps-year. Von Bortkiewicz noticed that this count rate follows quite precisely a Poisson distribution (Table 2-2, third column).

It is not surprising to see a rare occurrence of four horse-kick deaths in an army corps in one year. We can find from the Poisson distribution ($\mu = 0.70$ deaths per corps-year) that the probability of having at least four events in one corps-year is 0.0058. This looks rather small, but in 14 corps it gives the probability of 0.078 per year[6]. Hence, on average, we expect one such event in about 13 years (but see the 'Inter-arrival times' sub-section).

Inter-arrival times

This brings us to yet another interesting property of the Poisson distribution. How long do we have to wait between two

---

[5]The Law of Small Numbers.
[6]$1 - (1 - 0.0058)^{14} = 0.078$.

consecutive events (e.g. horse-kick deaths or radioactive decays)? Black dots in Figure 2-7a suggest that random events are distributed rather randomly. The time between two consecutive events, ΔT (called interarrival time), is a random variable with a known probability distribution. Its cumulative distribution is described by a very simple formula:

$$P(\Delta T < t) = 1 - e^{-\mu t}. \tag{2-14}$$

This is a direct consequence of equation (2-12), extended to an arbitrary time interval, t. The probability of having no events over time t is then $e^{-\mu t}$, and hence the probability of having *at least* one event over this interval is $1 - e^{-\mu t}$. This can be interpreted as the probability that the next event occurs within time t after the previous event. It might sound a bit weird, but the same formula applies to the waiting time after any arbitrarily chosen moment in time, not just the previous event. This is because the probability that an event occurs in a given period of time does not depend on whether another event just happened. The events are independent!

The mean of this distribution, called *mean inter-arrival time*, is $1/\mu$. This is how long, on average, we have to wait for the next event (but see below). The corresponding probability is $P(\Delta T < 1/\mu) = 1 - e^{-1} = 0.63$, and it doesn't depend on the actual value of the mean.

Figure 2-8 shows the distribution of inter-arrival times for the rare events of four horse-kick deaths in one corps-year. Obviously, the longer you wait, the higher the cumulative probability that this unfortunate event eventually happens. If you wait 20 years (the period over which the study was carried out), there is a cumulative ~79% probability of four deaths in a corps-year. After 60 years, the probability goes up to 99%. The longer you wait, the more likely it is that even an unlikely event will eventually happen. Perhaps Prussian army officials should not be surprised. Mind you, sometimes you might have to wait very long (see Exercise 2.3).

The mean inter-arrival time is $1/\mu = 13$ years in our example. This fact is often described, as 'the event happens (on average) once in 13 years'. To some extent, this is a true statement, but it can be very misleading. Random independent events are exactly what they are called: random and independent. The occurrence of an event does not influence another, or at least the influence is weak. Inter-arrival times vary greatly and certainly don't show any periodicity. For example, *on average*, there were just over a dozen of commercial airplane crashes a year in the past decade. But this doesn't mean they happen every month, and certainly doesn't imply that there is an air crash due, just because there was one

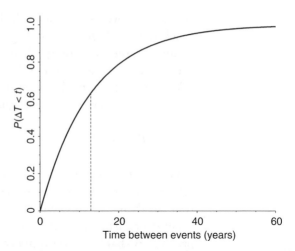

Figure 2-8. Cumulative distribution of inter-arrival times between events in a Poisson process, calculated for the mean event rate of $\mu = 0.078$ per year. The curve shows the probability of the next event occurring within t years after the previous event. The vertical dashed line shows the mean inter-arrival time of $\frac{1}{\mu} = 13$ years.

a month ago. Similarly, from the point of view of statistics, we should not be surprised if two of three major crashes occur in a single month. Random independent events do it sometimes.

Random independent events do not exhibit periodic behaviour.

2.9 Student's *t*-distribution

You will not find observable quantities in biology that obey a *t*-distribution. This distribution was created by mathematicians for statistical tests and estimating confidence intervals. I will discuss how this is done in Chapter 5. The first paper on *t*-distribution was published in 1908 by William Gosset, who worked at the Guinness Brewery in Dublin. He wasn't allowed by his employer to publish scientific papers, so he used a pseudonym, *Student*. This name was later popularized by the famous statistician Ronald Fisher.

Let us consider a population with mean μ, from which we draw a sample of size n. The sample's mean and standard deviation are M

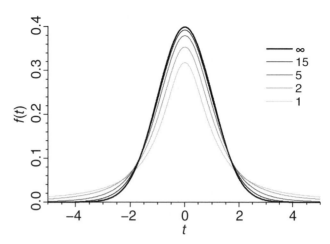

Figure 2-9. Student's *t*-distribution for an increasing number of degrees of freedom (shown in top-right corner). The thick black curve represents the infinite number of degrees of freedom, which corresponds to a Gaussian distribution $\mathcal{N}(0, 1)$.

and *SD*, respectively. I will explain all these concepts in Chapter 4. The quantity

$$t = \frac{M - \mu}{SD/\sqrt{n}} \tag{2-15}$$

See Section 4.8 for explanation of the degrees of freedom.

obeys the Student's *t*-distribution with $n - 1$ degrees of freedom. The number of degrees of freedom is defined as $v = n - 1$ because of the standard deviation in the denominator. This distribution is useful when you want to estimate uncertainty of the mean from repeated measurements (replicates). It is also used in comparing means from two samples with a *t*-test. I will show useful applications of the *t*-distribution later.

The Student's *t*-distribution is similar in shape to Gaussian distribution (see Figure 2-9). In particular, when the number of degrees of freedom is large, it actually approximates the Gaussian distribution. Note that, due to the definition of *t*, the distribution is centred on zero (so its mean is always zero). Its standard deviation is

$$\sigma = \sqrt{\frac{v}{v - 2}} \tag{2-16}$$

when $v > 2$. It is easy to notice that for a large number of points, $\sigma \approx 1$. This is when the *t*-distribution approximates the standardized Gaussian $\mathcal{N}(0, 1)$.

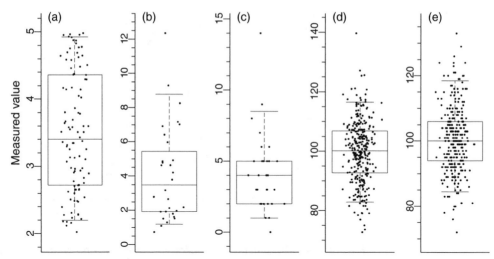

Figure 2-10. Each panel shows a sample drawn from an unknown probability distribution. Box plots show the median in the middle, each box stretches from the 25th to 75th percentiles and whiskers are from the 5th to 95th percentiles.

2.10 Exercises

Exercise 2.1
Look carefully at Figure 2-10. It shows five samples drawn from various probability distributions. For each sample, try to recognize the distribution and estimate its mean and standard deviation. Hint: pay attention to how data points are distributed. Which of these distributions are discrete, and which are continuous?

Exercise 2.2
In an experiment, you transfected a marker into a population of 3×10^5 cells. The marker functionally integrates into the genome at a rate of 1 in 10^5. What is the probability of obtaining at least one marked cell after this procedure?

Exercise 2.3
In the UK National Lottery, six balls are drawn out of 49. An easy calculation shows that there are

$$\binom{49}{6} = \frac{49!}{6!(49-6)!} = 13{,}983{,}816$$

possible ways of choosing six balls out of 49. Hence, the probability of winning the jackpot in one draw (matching all six balls) is $P_1 = 13{,}983{,}816^{-1} \approx 7.15 \times 10^{-8}$. If you play the lottery once every week, what is the mean winning rate (the mean number of jackpot wins per year)? What is the corresponding mean inter-arrival time between the wins (i.e. how long, on average, do you need to wait to win the jackpot)?

Exercise 2.4

Compare the Student's t-distribution with the Gaussian distribution. How well does the t-distribution approximate the Gaussian one, depending on the number of degrees of freedom? Hint: consider a particular tail probability, for example 5%, and compare tabulated cumulative distributions (available in this book).

Chapter 3

Measurement errors

If your experiment needs statistics, you ought to have done a better experiment.

—*Ernest Rutherford*

3.1 Where do errors come from?

The Latin word *error* means an error or mistake, or it means wandering or going astray. In everyday use when we say *error*, we usually mean *mistake*, which is an error caused by a fault: misjudgement, carelessness or forgetfulness. If you keep your cake in the oven for too long, because you misread the recipe or simply forgot about it, it will burn and this is a mistake. If you misjudge the distance to the oncoming traffic and overtake when you shouldn't, you might cause a serious accident; this is a potentially catastrophic mistake. In biology, we can talk about errors in DNA replication where a nucleotide is altered, which might lead to a mutation.

In statistics, the meaning of the word *error* is quite different. It refers to *measurement uncertainty*, the deviation of the measured quantity from its true value. In this meaning, an error is (usually) not a mistake. This is best demonstrated by repeated measurements, commonly called *replicates*. If you repeat the same experiment several times and measure the same quantity, you are likely to get a different result each time. In other words, you will observe *variability* in the measured value.

In statistics, errors refer to measurement uncertainties.

There are many potential sources of this variability. Consider the example from the beginning of Chapter 1: measuring gene

Understanding Statistical Error: A Primer for Biologists, First Edition. Marek Gierliński.
© 2016 John Wiley & Sons, Ltd. Published 2016 by John Wiley & Sons, Ltd.

Table 3-1. Examples of systematic and random errors.

Systematic	Random
Incorrect instrument calibration	Measurement errors
Model uncertainties	Sampling errors
Change in experimental conditions	Counting errors
Mistakes	Background noise
Sensitivity limits	Intrinsic variability

expression in a microarray experiment. The final measurement of the fluorescent dye emission is at the end of a long and complicated chain of processing, which introduces errors at each step. To complicate things even more, many subjects in biological measurements are intrinsically variable, and so is gene expression.

Systematic errors

Generally speaking, there are two types of uncertainties to consider: systematic and random errors (Table 3-1 and Figure 3-1). Sometimes these two categories overlap a little. Systematic errors are biases in measurements, typically caused by problems with instruments. They cause a shift or scaling of the mean of many replicates. For example, if the scale in a mercury thermometer is incorrectly placed, it will read temperatures consistently higher or lower by a constant offset. A digital thermometer translates resistance of a thermistor probe into temperature expressed, e.g., in degrees Celsius. If a multiplicative coefficient used for this conversion is incorrect (wrong calibration), you might end up with readouts shifted by an offset proportional to temperature. Increasingly, results in modern biology are obtained through complex mathematical models implemented in computer software (e.g. mass spectrometry). These models use various approximations of underlying physics with inherent inaccuracy, and can be a source of systematic errors.

If you change experimental conditions, you are likely to obtain different results, again with the mean value systematically shifted. For example, many experiments are sensitive to temperature, so either setting the wrong temperature or allowing for temperature drift will alter your measurements. All these systematic errors typically originate in a flawed experimental setup and usually can be eliminated (or at least minimized) in well-designed experiments.

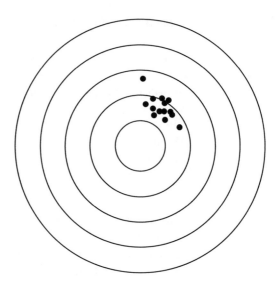

Figure 3-1. Intuitive illustration of systematic and random errors. The target symbolizes the true value of the measured quantity; the black dots represent a series of measurements. Systematic errors shift all measurements in the same direction, changing their mean value. Random errors cause scatter (variation) in measurements, but do not affect their mean.

Systematic errors can be minimized in good experiments.

I am not going to discuss systematic errors anymore in this book. They belong in the world of good experimental practice, following protocols and instruction manuals. They have little to do with statistics, and statistics is what I'm trying to explain here.

Random errors

Random errors, as the name suggests, change measurements in a random way. They have random direction and random magnitude, and should average to zero over many measurements. Therefore, random errors do not introduce systematic bias into results and do not affect the mean of many replicates. At least in an ideal world.

Random errors are inherent to every experiment in biology, and while they may in some cases be reduced, they cannot be removed simply by good experimental practice. You can estimate and reduce random errors by taking multiple measurements (replicates) as many times as possible.

You cannot eliminate random errors; you have to live with them.

In Sections 3.3 through 3.5, I will discuss different types of random errors. This particular division or classification is entirely arbitrary, and you might find different definitions in other textbooks. It merely reflects what I want to tell you about random errors and some of their properties.

3.2 Simple model of random measurement errors

Consider the following experiment. Let us determine the strength of oxalic acid in a sample. The method we are going to use is a titration, where we find the volume of sodium hydroxide (NaOH) solution required to neutralize a given volume of the acid by observing the change in colour of a phenolphthalein indicator. This procedure consists of several steps of volume and weight measurements, each of them with a certain reading error. All these errors will contribute to the final result. Let's have a look at uncertainties involved in this experiment:

- Volume of the acid sample
- Volume of NaOH solution used at this point
- Accuracy of NaOH concentration, which is affected by
  - Weight of solid NaOH dissolved
  - Volume of water added
- Judgement of the indicator colour.

This is a very simple example. In real life, many of the contributing errors will be hidden from the observer. This happens, in particular, when raw data from a complicated instrument are processed by computer software, for example in mass spectrometry or microarray experiments. There is no way of knowing all of the possible uncertainties arising between the observation and the result.

But we shall not despair. We can at least try to understand how such errors affect the final result. In order to do this, let us build a very simple model of measurement errors. Consider a measurement of a quantity, whose true (unknown to us) value is m_0. As in the example above, the measurement is perturbed by several small uncertainties. Each of them contributes a small random deviation, ε_i, which can be either positive or negative. The measured value is then

$$m = m_0 + \sum_i \varepsilon_i. \tag{3-1}$$

The central limit theorem is described in Section 2.5. Mathematically speaking, each perturbation ε_i can be regarded as a random variable of mean zero, small standard deviation and unknown distribution. Now, we can use the central limit theorem, which states that the sum of many random variables with arbitrary distributions is Gaussian. *Ergo*, the measured value, m, is normally distributed around m_0. It is easy to notice that contributions ε_i with non-zero mean would be responsible for a systematic shift in the measurement.

This simple model of errors was first introduced by Laplace in 1783. It is illustrated in Figure 3-2. Let us assume, for even more simplicity, that all our perturbations are the same in magnitude and can take only two values: $+\varepsilon$ or $-\varepsilon$, with equal probability of 1/2. You can model this process by tossing a coin at each step to decide the sign of the perturbation. We start with the unperturbed, true value m_0 at the top of the plot in Figure 3-2a. After the first coin flip, we can have the first perturbation of $+\varepsilon$ or $-\varepsilon$, with equal probability of 1/2. After the second coin toss, we have four possible outcomes, all of them with equal probability of 1/4:

$$-\varepsilon - \varepsilon = -2\varepsilon$$
$$-\varepsilon + \varepsilon = 0$$
$$+\varepsilon - \varepsilon = 0$$
$$+\varepsilon + \varepsilon = 2\varepsilon$$

The second and third combination of epsilons lead to the same deviation, so this result is twice more likely than the first and fourth combinations. Hence, the probabilities of obtaining -2ε, 0 and 2ε are 1/4, 1/2 and 1/4, respectively. This is illustrated in the second row of probabilities in Figure 3-2a. If we continue further, we will find a wider range of sums of perturbations, with values near zero more likely than values further away. This forms a familiar binomial distribution and is shown in Figure 3-2b. For a large number of perturbations, it approximates the Gaussian distribution.

The binomial distribution is explained in Section 2.7.

Random measurement errors are normally distributed.

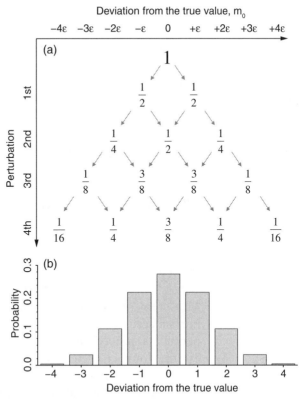

Figure 3-2. Simple model of random measurement errors. The horizontal axis shows the deviation from the true, unknown value of the measured quantity. There are several perturbations from the true value in the process of measurement, each of them $+\varepsilon$ or $-\varepsilon$, with equal probabilities. The measured result is the true value plus the sum of all perturbations. Panel (a) shows probabilities of obtaining a given deviation. There are more paths leading to small than large deviations, so small deviations are more likely. They form a binomial distribution, an example of which is shown in panel (b). For a large number of perturbations, it becomes Gaussian.

But what does it mean? As usually happens in statistics, theoretical distributions can be translated into real frequency counts, when a measurement is repeated. The error model described here suggests that if you were to measure a given quantity many, many times, these measurements would form a Gaussian distribution with the mean equal to the true value of the measured quantity. Of course, in the real world the number of possible measurements (replicates) is limited, and their mean is only an estimator of the true value.

Statistical estimators will be discussed in Chapter 4.

Another consequence of the stochastic nature of errors is that a quoted error is a statement of probability. A number and its error written as '12 ± 3' does not mean that the measured value is between 9 and 15 for sure. It only means that we have some *confidence* that the true measured quantity is in this range, but we cannot guarantee this. Instead, we can produce a probability that the given range contains the quantity in question. But, despite our best efforts, the true measured quantity might be outside the quoted region. Tough luck.

I will explain confidence intervals in Chapter 5.

> Reported error is always a statement of probability.

3.3   Intrinsic variability

Measurements are uncertain because the very process of measuring is inaccurate. By building better instruments and refining experimental procedures, we might reduce this uncertainty to some extent. However, even if we managed, by some miracle, to reduce random and systematic measurement errors down to negligible values, there still would be an uncertainty (often considerable) in an outcome of a typical biological experiment. This is due to the fact that the subject of measurements varies by its nature. Such variability can be revealed in repeated measurements.

Imagine you want to find the height of a person. With a sufficiently accurate height measure, you can probably establish how tall someone is with an associated error of less than 1 cm. If you repeat the measurement many times, the majority of your results should be within ±0.5 cm of the true height. This is fine, but it only tells you about the height of one particular person. What if you want to know what a *typical* human height is? You have no guarantee that your subject is of average stature (and, almost certainly, he or she is not). You can gather a group of people, for example your work colleagues, and measure them. Certainly, each of them is going to have a different height, and the *scatter* between measurements is going to be much larger than our measurement error of ±0.5 cm.

This is a considerable problem. Not only are measurements inaccurate, but also the things that are measured change their properties in a random fashion. This intrinsic variability makes every measurement different. To make things worse, the extent of this variability in biological systems is usually larger than any other type of error.

Typically, intrinsic biological variability is the dominant source of uncertainty.

3.4 Sampling error

Can we eliminate the intrinsic variability by taking multiple measurements of the same thing? After all, if the measured thing varies, we are probably interested in its average behaviour. For example, from a sample of height measurements (your work colleagues), we can infer the mean height in the population of humans. This is fine, but if you select another group of people (e.g. your family), you will find a different set of heights and a different mean. Height is different from person to person; mean height is different from sample to sample. This is the same problem again: repeated measurements (a sample) can give us an estimate of the mean height, but this estimate is inaccurate in its own way. Drat!

Sample and population are discussed in Section 4.1; statistical estimators are the topic of Chapter 4.

Selecting a sample from the population introduces a new type of uncertainty, called the *sampling error*. A sample only approximates the population and produces *estimators* of true population parameters. Luckily, with proper statistical tools we can assess uncertainties of these estimators. I will explain these concepts later in the book. For the time being, let us avail ourselves of the following (rather vague, I know) definition:

Sampling error is caused by observing a sample instead of the entire population.

In this example, measurement errors are negligible in comparison with sample scatter. Although this is rather typical in biology, it is not always true, and the spread in the results can be due to measurement error as well. Quite often, we don't know where the observed variability arises: it could be an intangible mixture of both effects.

Sampling in time

Sampling can be performed not only by measuring several subjects but also by measuring one subject at different time points. Remarkably, these two types of sampling (across subjects and across time) under certain conditions can give very similar results. The main assumptions are that the observed process is *stationary* – its properties (mean, standard deviation etc.) do not change in time – and that all subjects vary in the same way. In mathematics, this property is described by the *ergodic theorem*. This applies to many physical systems, as laws of physics do not change from one laboratory to another. An electron is exactly the same in your lab and on the Moon. In biology, however, we cannot assume that all subjects are equal. Even 'identical' twins might respond differently to the same stimulus. Therefore, you should exercise the utmost caution when trying to extend results from one subject to the entire population.

Ergodic theorem: under certain conditions sampling in space can be replaced with sampling in time.

More about replication and pseudoreplication in Section 5.11.

Repeated measurements from the same subject are not replicates.

Obviously, when you measure human height, you don't expect it to vary in time for one person[1]. On the other hand, if you want to find a range of body temperatures for a healthy person, you can measure one individual for several days, or measure a group of people once. If they are all 'typical' and healthy, you should get similar results, but this is not guaranteed. From the point of view of an experimental biologist, it is *always* better to have a sample of subjects (replicates), because any one subject might not be representative. After stating these caveats, I can say that sampling in time (e.g. from the same organism) is a valid scientific method, as long as you know what you are doing.

Consider measuring protein abundance *in vivo*, estimated by the intensity of a fluorescent marker. We assume that the expression level of this protein is stable. Like many other processes in the cell, protein abundance is stochastic and varies at all timescales. The grey curve in Figure 3-3 shows how such a process might appear for one protein, although this is only a simulation done for illustrative purposes. The true nature of this variability is unknown, as we can only measure it at a certain time step, 60 seconds in this example. If the process is stationary, the true unknown intensity (grey curve) can be regarded as a population, whereas timed measurements (black dots) can be treated as a sample taken from this population. From the sample you can only find some

Population, sample and statistical estimators are explained in Chapter 4.

---

[1]Strictly speaking, height can and does vary for an individual as a result of spinal compression, even during a day. You are taller in the morning than in the evening!

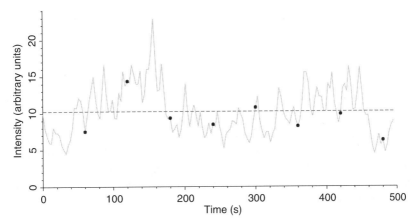

Figure 3-3. Example of a random stationary process where intensity is measured every 60 seconds. The grey line represents intrinsic, unobserved variability; the black dots show actual measurements. These can be regarded as population and sample, respectively. The true (unknown) mean intensity is 10.2, shown by the dashed line. The mean of the measurements is 9.4, and the standard deviation is 2.5. These are not real data but a simulation.

statistical estimators, like mean and standard deviation, that approximate true unknown values. This kind of sampling in time doesn't apply to experiments, where the subjects change over time.

Taking a sample either from a population of subjects or from a time series might look pretty straightforward, but the implications are far-reaching and not simple at all. In experimental biology, sampling (having several replicates) is the main and sometimes the only way of estimating measurement uncertainties.

3.5   Simple measurement errors

Measurement errors are random in nature and typically require replicated experiments to estimate their size. From repeated experiments, we can estimate the extent of variability resulting from both biology and technical protocol. There are, however, a few types of errors that can be estimated directly from measurements without the need of replicates. The uncertainties I'm going to discuss in this section are purely measurement errors, and they do not take into account the biological variability you encounter in most experiments. Therefore, even if you can estimate measurement errors directly, I still recommend doing the experiment in replicates.

Reading error

Direct measurements typically involve reading either a scale (e.g. a ruler or analogue voltmeter) or a digital display (e.g. a digital

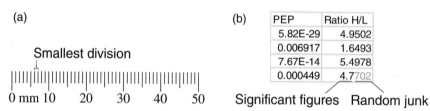

(a)

Smallest division

0 mm 10   20   30   40   50

(b)

PEP	Ratio H/L
5.82E-29	4.9502
0.006917	1.6493
7.67E-14	5.4978
0.000449	4.7702

Significant figures Random junk

Figure 3-4. Reading errors. (a) Reading error of an analogue scale (e.g. a ruler) is typically half of the smallest division, but there might be other factors contributing. (b) Fragment of a table created by protein mass spectrometry analysis software. Digital readouts, in particular when data are uploaded into a computer, are often shown with more figures than the actual resolution. This is OK for performing further calculations, but non-significant figures have to be rejected when presenting data in a publication.

voltmeter or computer). Measuring equipment is limited by its own resolution. For example, if you use a ruler with a millimetre scale (Figure 3-4a), you should be able to read the length to the nearest millimetre, at best. If you can't, you should probably visit your optician. The reasonable estimate of uncertainty of length would be, in this case, $\Delta l = 0.5$ mm. We use half of the smallest division because error is typically represented as $\pm \Delta l$, one error up and one error down; the total size of error is $2\Delta l$. This applies to any device with a scale.

> The reading error of a scale with markings is half of the smallest division.

In real life, reading error is often larger than this, as there is always a human involved and *errare humanum est*. A measurement from a scale will vary depending on how your eye is lined up with the subject and the reading device (e.g. a ruler). If your measured subject is not placed immediately behind the ruler, the readout will increase and decrease as you move your head left and right due to parallax. Sometimes, it might be difficult to point where your subject begins and ends while measuring its length. If you have ever tried to measure the length of a mouse's tail, you know what I'm talking about. Occasionally, a scale might be too fine to read accurately even by a person with good eyes.

> Only the observer can ultimately decide what the actual reading error is.

I should point out that, quite often, the scale reading error is largely irrelevant if the subject-to-subject variation (in whatever you measure) is much larger than the reading error.

Digital readouts are an entirely different matter. You might be tempted to assume that the reading error is half of the last digit displayed. If a digital thermometer shows 36.6 °C, the reading error might be ± 0.05 °C. Or it might not. You really need to consult the specifications of the instrument. For example, one particular high-precision digital thermometer has, according to the manufacturer, 'resolution of 0.01 °C' and 'accuracy better than 0.04 °C'. The first number refers to the number of decimal places in the digital readout, and the second is the actual measurement error.

All becomes even more obscure when it comes to instruments transferring data into a computer. Computers often present numbers with lots of digits (Figure 3-4b), but don't be fooled into thinking that actual measurements are so precise. These numbers are usually a result of several calculations (sometimes, a complex chain of sophisticated processing by very expensive software) and are stored with the precision of a computer number. There will be *I will explain significant* lots of figures in the number, but only a few of them are significant. *figures in Section 6.4.* The rest are simply random junk.

> Beware of superfluous precision in digital readouts.

There is a reason why raw instrumental data are stored with superfluous precision. Typically, these numbers are used for further calculations; for example, mean and confidence interval might be computed from repeated measurements. If we truncate numbers down to their actual precision at the beginning of calculations, we might lose some of this precision in processing steps due to further rounding and truncation. It is better to use the redundant digits *See Section 6.4 for* to make sure that rounding errors do not contribute to our final *quoting numbers with* result. When this is done, the final number and its error should be *errors.* truncated according to their actual precision.

Counting error

Consider an experiment in which we perform 10^{-5} dilution plating of some bacteria. The source culture is diluted with sterile liquid in subsequent steps and transferred onto an agar plate. We let the bacteria grow, and after a while we find 11 colonies. We assume that each colony originated from just one bacterium from the diluted culture. From this, it is easy to work out the bacterial count in the source culture of 1.1×10^6, but what is the error of this estimate? As usual, the best way of finding errors is to perform replicates of your experiment and find the mean and the standard error (see Chapter 4). However, in this particular case, there is a simpler and quicker way.

I discuss the Poisson distribution in Section 2.8.
Rare, random, independent events obey a Poisson law. Bacterial colonies on the plate appear randomly, they are not too frequent if diluted enough, and each colony should be independent of another if the experimental conditions are good. Under these assumptions, the count of bacterial colonies on the plate is Poissonian. There-

See equation (2-13).
fore, the mean count equals its variance. We can accept the observed count as an estimate of the mean, and we use its square root (standard deviation) as an estimate of uncertainty. As we have found 11 colonies, the error is then $\sqrt{11} \approx 3$, which, after scaling due to dilution, gives $(1.1 \pm 0.3) \times 10^6$ bacteria in the culture.

> ## Counting error can be estimated by the square root of the count.

I must stress that counted objects or events must be independent to conform to the Poisson distribution. If cells attract or repel each other, their distribution can be non-Poissonian, and then counting error does not follow this simple rule.

Let us try to understand more clearly what this error estimate really means. First of all, the fact that we have 11 colonies does *not* mean that the sample size is 11. The sample size is one, as we have one measurement and the result of this measurement is a number, 11. This can be considered as one realization of a random variable. The probability distribution of this variable can be revealed in repeated experiments. If we managed to perform the entire bacterial colony experiment many times (e.g. 10,000), we could plot the underlying probability distribution and measure its parameters. As this would be rather impractical, I have simulated such an experiment in a computer. We can call it a *thought experiment*, or *gedankenexperiment*, as physicists often call it. Bacterial colony counts from the first 16 plates, together with their estimated counting errors (square root of the count), are shown in

The rules of quoting numbers with errors are described in Section 6.4.
Figure 3-5a. The first plate, with 11 ± 3 counts, is from our original experiment. Figure 3-5b shows the distribution of counts from all 10,000 plates. This approximates a Poisson distribution. The point and error bar above the distribution represent the mean ±

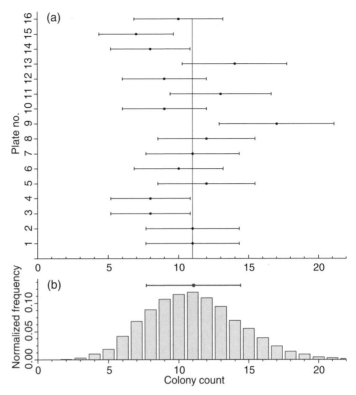

Figure 3-5. A thought experiment in which bacterial colonies are counted on agar plates. The experiment is repeated 10,000 (yes!) times. (a) Counts from the first 16 plates. Dots show the actual count, c_i, from plate i, and error bars are Poissonian, $\pm\sqrt{c_i}$. The vertical line represents the population mean, $\mu = 11$ counts per plate. (b) Distribution of colony counts from all 10,000 plates. The point in the middle and error bar show the mean and the standard deviation from the entire distribution.

standard deviation of all data, which were 11.06 ± 3.34 in this particular simulation. This is very well matched by our first plate, shown at the bottom of Figure 3-5a. However, there is absolutely no guarantee that another measurement is going to give us anything so close to the true mean! We are not going to be so lucky every time. For example, there were 17 colonies on plate 9, and our error estimate was $\sqrt{17} \approx 4$. Hence, the true mean of $\mu = 11$ is outside the error bar, 17 ± 4. In fact, about one-third of all 10,000 measurements do not have the true mean within their error bars. This demonstrates the probabilistic nature of error bars.

Standard error of the mean is explained in Section 4.5; confidence intervals for count data are discussed in Section 5.9.

Counting error defined as the square root of the count is a *standard error*. It is a bit similar to the standard error of the mean in a way that estimates the width of the sampling distribution (more about this in Section 4.5). But it is not a confidence interval. To find a proper confidence interval on a count, we need to do a few mathematical tricks. I will explain this in Chapter 5.

3.6 Exercises

Exercise 3.1
Measure the length of this book using a ruler. What is the measurement error? If you had many copies of this book and measured each of them, what could you tell about the book length and its error?

Exercise 3.2
A local newspaper reported that Springfield is 'the murder capital of the country'. There were six murders in Springfield and 19 in the notorious Capitol City in the last year, reported by the police. However, when the size of each city is taken into account, there were 4.1 and 3.2 murders per 100,000 population in Springfield and Capitol City, respectively. The newspaper concluded that Springfield has a higher murder rate and therefore deserves the title of 'murder capital'.

Is this title justified? Estimate uncertainties on murder rates in both cities and compare them. Hint: if you multiply a number by a constant, its error scales by the same constant.

Chapter 4

Statistical estimators

The average human has one breast and one testicle.

—*Des MacHale*

Consider a very simple experiment in which we weighed five mice. Using electronic scales with 0.01 g accuracy, we found the following weights: 21.69, 25.00, 11.68, 17.05 and 18.61 g. From these numbers, we can find the mean weight of 18.8 g with standard deviation of 5.0 g and standard error of 2.2 g. Or, we can find the median of 18.6 g. All these numbers are examples of *statistical estimators*, and they describe sample properties. If you are an experimental biologist, you have probably calculated hundreds and thousands of statistical estimators without even realizing what they were called, just like Mr Jourdain was surprised that he had been speaking prose all those years. In this chapter, I will explain and discuss the most commonly used statistical estimators. But firstly, we have to talk about the meaning of population and sample.

4.1 Population and sample

The terms *population* and *sample* are nicked from social sciences, where they are used in a very literal sense. A population is a large group of people, for example the population of a country. A sample is a small group selected, for example, either for polls or for some other type of interrogation. It should be small enough to be easily manageable. It is rather impractical to send a questionnaire to everybody in the country, so that is done only during the census once in 10 or so years. However, a sample should be large enough to give the required statistical confidence in the result. Asking your auntie about her political views does not necessarily reflect the views of the entire population. A typical opinion poll involves about a thousand people.

Understanding Statistical Error: A Primer for Biologists, First Edition. Marek Gierliński.
© 2016 John Wiley & Sons, Ltd. Published 2016 by John Wiley & Sons, Ltd.

Table 4-1. Population and sample: a summary.

Population	Sample
Population can be a somewhat abstract concept.	Sample is what you get from your experiments.
All possible specimens or measurements	A representative selection
Huge size, impossible to handle	Manageable size, n measurements
Examples: • all possible mice (that lived in the past and will live in the future) • all people with eczema • all possible measurements of gene expression (infinite population)	Examples: • 12 mice in a particular experiment • 26 patients with eczema • 5 biological replicates to measure gene expression
Population is described by an unknown distribution with mean μ, and standard deviation σ.	From a sample we can calculate *statistical estimators* of population's μ and σ

The sample should be representative. In other words, it should reflect the distribution of important characteristics of the population: gender, age, education, income and so on. Querying Premier League footballers about their cars is not the best way of finding what an average person drives.

In statistics, population and sample are slightly redefined, although the main idea remains the same (Table 4-1). In an experiment, you select a subset of all possible measurements or values. The sample is usually easy to define; this is what you get from your experiment. Typically, it is a set of numbers representing properties or characteristics of interest. Population can be a bit vague. If you want to study mice, the population might represent all mice in a given geographical region, all mice on Earth or even all mice that ever existed. It gets even murkier when you measure more abstract quantities, such as gene expression. The sample is your spreadsheet with data, but what about the population? It is an abstract set of all possible gene expression levels. Imagine you have an infinite amount of time and infinite funds. You repeat your experiment in infinite replicates and collect an infinite amount of gene expression levels. That would be a population. But you don't have infinite resources, so you can't do this.

Population parameters characterizing a probability distribution are discussed in Section 2.3.

From the point of view of an experimental biologist, a population has certain characteristics and we try to understand them by studying a representative sample. The population is often described by a theoretical probability distribution, for example a Gaussian or Poisson distribution. These distributions are characterized by (unknown) parameters, such as mean, μ, and standard deviation, σ.

> Population mean and standard deviation are unknown and can only be estimated.

It is often impractical to study an entire population, even if it is finite in size. Instead, we study a sample that is hopefully representative of the population and find its mean and standard deviation. It is imperative to stress that these are not identical with the population parameters. They only *approximate* them. Sample characteristics are only estimators.

Please note that the term *sample* when used in statistics has a different meaning from what biologists are used to. In experimental biology, *sample* usually means *specimen*, something you are going to study. If you prepare your cell culture in five plates and send them for protein quantification, there will be five biological samples to be analysed. However, when these specimens are processed and proteins are quantified, they will make (for a given protein) one sample of protein abundances with size $n = 5$. In this meaning, a sample is a set of numbers, usually obtained from biological replicates. I will use the term *sample* in its statistical meaning.

> In statistics, a sample is a set of numbers, usually obtained from quantification of biological replicates.

4.2   What is a statistical estimator?

In the sixteenth-century surveyor's manual *Geometrei*, Jacob Köbel (Köbel 1535) defines a unit of length (see Figure 4-1):

> Stand at the door of a church on a Sunday and bid 16 men to stop, tall ones and small ones, as they happen to pass out when the service is finished; then make them put their left feet one behind the other, and

Figure 4-1. 'Right and lawful rood' from *Geometrei*, by Jacob Köbel (Frankfurt 1535). A sixteenth-century unit of measure estimation by random sampling from a local population.

the length thus obtained shall be a right and lawful rood[1] to measure and survey the land with, and the 16th part of it shall be the right and lawful foot.

This is quite a remarkable instruction. Almost 500 years ago, Köbel introduced representative random sampling from a population and standardized units of measure. He used 16 'replicates' to minimize random error. Finally, he calculated the sample mean – a statistical estimator of the population mean.

Nowadays we do exactly the same. We take a sample from the population to study its unknown properties. Typically we would calculate sample mean and standard deviation. To distinguish them from population parameters μ and σ, I will denote them as M and SD, respectively. They are called *statistical estimators*.

---

[1]*Rood* was a unit of measure equal to about 16 feet.

A statistical estimator is a quantity derived from the sample to estimate a parameter of the population.

An estimator from the sample is usually different, but hopefully not hugely different, from the population parameter. Consider a Gaussian mouse population with a mean body weight of 20 g and a standard deviation of 5 g. I have just made up these numbers, as I cannot be bothered to weigh all mice in the known universe. From this population we select a sample of 30 mice, for example by breeding them in the laboratory. We can weigh them easily, and the distribution of their weights is shown in Figure 4-2 together with the Gaussian population distribution. The mean from this sample is very close to the true mean, but it doesn't have to be. So, how good is our sample mean? It is crucial to know the 'quality' of a statistical estimator, as otherwise it would be useless. In other

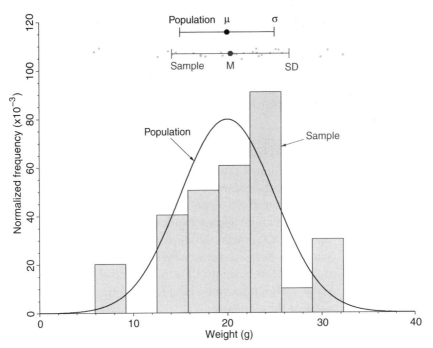

Figure 4-2. Population and sample of mouse body weight. Solid curve represents the (hypothetical) distribution of weights in mouse population. The black circle with error bars at the top of the figure represents $\mu \pm \sigma$. For the purpose of this example, the population was assumed to be Gaussian with $\mu = 20$ g and $\sigma = 5$ g. The sample of 30 mice is shown in grey points, with sample mean ($M = 20.4$ g) and standard deviation ($SD = 6.2$ g) overlaid. Shaded bars show the histogram of sample distribution.

words, from our sample mean and standard deviation, we have to come up with a robust *error of the mean*. We want to be able to say with a certain level of confidence that the population mean is within a particular range of values. I will show you how to do this in Chapter 5.

4.3   Estimator bias

In an ideal world, an estimator should represent the true value of the parameter estimated as accurately as possible. Obviously, by selecting a (random) sample, a sampling error is introduced and the estimator deviates from the true value. You can't really do anything about it except take a very large sample in order to minimize the deviation. Such is the nature of things.

Since we can't beat statistics and get rid of this deviation, we can at least try to see how bad it is. In particular, it would be nice to know whether the deviation is either random or systematic. A random deviation should be negligible on average, whereas a systematic deviation would introduce... a systematic shift in the estimator.

A systematic deviation between the estimator and the true parameter is called *bias*.

Let's go back to our mice. Every time you take a sample of 30 mice, the mean sample body weight, M, is going to be different. However, the mean value of M over many samples is going to be very close to the true population mean, μ. If you extend this sampling into an imaginary experiment, in which you take an infinite number of samples, the mean M will be exactly equal to μ. The sample mean is an *unbiased* estimator of μ. Analysing an infinite number of mice might be a bit beyond the abilities of a modestly funded laboratory, but what I have in mind is a theoretical, *expected value* of the estimator.

The expected value of an unbiased estimator equals the true parameter.

To demonstrate the bias with reasonable precision, one can do a computer simulation. I will show such an example later in this chapter. Or, in many cases, it is possible to perform an exact mathematical derivation to find the bias. For example, it turns out that the arithmetic mean is unbiased, which is very nice. Standard deviation, on the other hand, is biased (especially in small samples), although this bias can be corrected (see the 'Unbiased estimator of standard deviation' sub-section).

4.4 Commonly used statistical estimators

Mean

The most commonly used statistical estimator is the sample mean, M. For a sample of measurements $x_1, x_2, \ldots, x_n$, their *arithmetic mean* is defined as

$$M = \frac{1}{n} \sum_{i=1}^{n} x_i. \tag{4-1}$$

The error of the sample mean can be estimated by the standard error of the mean, although you might want to use confidence intervals to give it some statistical meaning.

Standard error of the mean and confidence intervals are described in Sections 4.5 and 5.4, respectively.

Please note that the sample mean, M, is markedly different from the random variable mean, μ, as defined by equations (2-3) and (2-4). From a slightly simplified point of view, the population and its probability distribution can be described by a random variable, whereas the sample is a set of realizations of this random variable. Hence, the random variable mean is the population mean, and the sample mean approximates it. This holds true for other estimators described in this section.

The sample mean has one interesting property. Deviations of individual measurements from the mean, $x_i - M$, are called *residuals*. The sum of all residuals is zero:

$$\sum_{i=1}^{n} (x_i - M) = \sum_{i=1}^{n} x_i - nM = \sum_{i=1}^{n} x_i - \sum_{i=1}^{n} x_i = 0. \tag{4-2}$$

In particular, we can state the following:

The mean residual is always zero: $\langle x - M \rangle = 0$.

Weighted mean

Occasionally, when data points have different importance, or *weight*, we might need to calculate a weighted mean. If a measurement x_i has a weight w_i, the weighted mean is described by the following formula:

$$M_w = \frac{\sum_{i=1}^{n} w_i x_i}{\sum_{i=1}^{n} w_i}, \tag{4-3}$$

If all weights are equal, this formula reduces to the arithmetic mean [equation (4-1)].

The weighted mean is usually used to average points with different errors. For example, if we were asked to find a mean of two numbers, 3 and 7, then the result would be 5. However, if these numbers had very different errors, 3 ± 0.5 and 7 ± 6, then we would assign more weight or importance to the first number, as it has much smaller error. The mean of these two points should be closer to 3 than to 7. This can be achieved by using the following weights:

$$w_i = \frac{1}{SE_i^2},$$

See Section 4.5 for standard errors. where SE_i is the standard error of x_i. In fact, other types of error estimates can be used as well. The weighted mean is then

$$M_w = \frac{\sum_{i=1}^{n} \frac{x_i}{SE_i^2}}{\sum_{i=1}^{n} \frac{1}{SE_i^2}}. \tag{4-4}$$

Standard error of the weighted mean is shown in Section 4.6. This particular definition has some desired statistical properties. For example, the error of the weighted mean can be derived directly from equation (4-4) (more about this later in this chapter).

The difference between the 'normal' unweighted arithmetic mean and the mean weighted by errors can be demonstrated in the following example. Imagine an experiment in which we count events from a 'source' and a 'background'. These could be any types of counts: for example, the number of cells under two conditions, or the number of sequenced reads mapped against a particular region of the reference genome. Now, we want to find a source-to-background ratio and its mean value. An example of such data is shown in Figure 4-3.

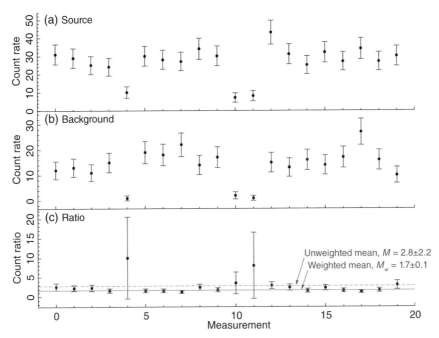

Figure 4-3. Illustration of the weighted mean. The data come from a simulated counting experiment. Panels (a) and (b) show source and background count rates, respectively. The errors bars represent counting errors. Panel (c) shows the source-to-background ratio with propagated errors (see Chapter 7). The horizontal solid and dashed lines show the weighted and unweighted mean, respectively. The errors of the unweighted and weighted mean are their corresponding standard errors.

Error propagation for a ratio is discussed in Section 7.3.

As it happens in real experiments, for some reason, the count rate occasionally drops, creating spurious high count ratios. However, as we propagated errors correctly from counts into count ratios, these low-count data points have large errors (see Figure 4-3c). The unweighted mean doesn't know about error bars and is 'dragged up' by the two measurements with high ratios. You could argue that these are outliers and should be removed from the sample manually. This, however, is data manipulation bordering on cheating. The weighted mean, on the other hand, assigns very low importance to data with large errors and effectively ignores them, so you don't have to worry about them.

Geometric mean

The geometric mean is similar is some way to the arithmetic mean, except that numbers are not added but are multiplied, and then the

nth root is taken:

$$G = \left(\prod_{i=1}^{n} x_i \right)^{\frac{1}{n}} = \sqrt[n]{x_1 \times x_2 \times \ldots \times x_n}, \tag{4-5}$$

Log-normal distribution is discussed in Section 2.6.   The geometric mean can be used in cases where data points (measurements) are meant to be multiplied rather than added. For example, if your data are log-normally distributed, you might want to use the geometric mean to describe its central value. However, in my opinion, the geometric mean is redundant and there is no need to use it. It is easy to show that the 'ordinary' arithmetic mean of logarithms is equal to the logarithm of the geometric mean,

$$M_{log} = \frac{1}{n} \sum_{i=1}^{n} \log x_i = \frac{1}{n} \log \prod_{i=1}^{n} x_i = \log \left(\prod_{i=1}^{n} x_i \right)^{\frac{1}{n}} = \log G. \tag{4-6}$$

This is because the sum of logarithms is the logarithm of the product. Hence, instead of using the awkward geometric mean, you should log-transform your data (i.e. replace x_i with $\log x_i$, for each i). Then, you can use the bog-standard arithmetic mean. The advantage of logarithmic data is that the standard deviation works properly. If your data are log-normally distributed, the standard deviation doesn't make much sense, as demonstrated in Section 2.6.

> Don't use the geometric mean; log-transform your data instead.

Median

The median is a cousin of the mean. Likewise, it represents the central location of the data. It splits the sample into two halves, lower and upper, each with an equal number of data points. All data points in the lower part are smaller than (or equal to) the points in the upper half. I will denote the sample median as $\widetilde{M}$. The population median, θ, divides the population in two halves: $P(X \geq \theta) = P(X \leq \theta) = \frac{1}{2}$. As usual, $\widetilde{M}$ is the estimator of θ.

In order to find the median of a sample $x_1, x_2, \ldots, x_n$, let us first sort our sample in ascending order, $x_{(1)} \leq x_{(2)} \leq \ldots \leq x_{(n)}$. Indices in brackets indicate the rank in the ordered sample, from the sample minimum, $x_{(1)}$, to the sample maximum, $x_{(n)}$. The sample median sits in the middle of this sequence. If n is odd, then $\widetilde{M} = x_{(\frac{n+1}{2})}$. For example, for $n = 17$, the median is the ninth point in the sorted sample, $\widetilde{M} = x_{(9)}$. If n is even, the sample median is in the middle of the two central points: $\widetilde{M} = \frac{1}{2}(x_{(\frac{n}{2})} + x_{(\frac{n}{2}+1)})$. For example, for $n = 18$, the median is between the ninth and 10th points in the sorted sample: $\widetilde{M} = \frac{1}{2}(x_{(9)} + x_{(10)})$. In practice, there are more efficient algorithms for finding the median without the need of sorting.

The difference between the mean and the median is that the mean takes data values into account, whereas the median is derived from their ranking. This makes the median immune to extreme outliers. If you take a sample of five data points and sort them, the median will be always equal to the third point, regardless of how small or large the extremes are. The mean, on the other hand, can be significantly altered by a small number of outliers, as demonstrated in Figure 4-3c.

> The median is not sensitive to outliers.

Another useful property of the median is that it gives a better 'feeling' of the sample's (or population's) central value when the distribution is very skewed. A classic real-life example is the distribution of salaries, which is very asymmetric, with a small fraction of the population earning a lot. According to the Office for National Statistics, the mean gross weekly salary in the United Kingdom in 2012 was £607, whereas the median was £506.

> The median represents the middle of the data in skewed distributions.

Standard deviation

Mean (or median) represents the central location of the data, but it doesn't tell you anything about the spread of the sample. Sometimes measurements can be near-identical, sitting tightly close to their mean value. In other experiments data points can be scattered widely and wildly all over the place. Surely, we would rather trust the mean when data are not too variable across the sample. We could use the sample's variability or spread to estimate the confidence of the mean.

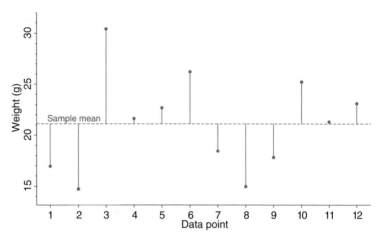

Figure 4-4. Variability (or spread) in data. The graph illustrates 12 data points (dots) and their mean, $M = 21.1$ g (horizontal dashed line). Vertical lines show deviations from the mean (residuals). The standard and mean deviations are $SD_{n-1} = 4.8$ g and $MD = 3.8$ g, respectively.

Imagine you have a sample of n data points: $x_1, x_2, \ldots, x_n$. An example is shown in Figure 4-4. The (arithmetic) mean, M, of these data is described by equation (4-1). We could estimate the variability in the data by looking at how far they deviate from the mean. For this, we can calculate individual residuals, $x_i - M$. If there is little variability in the data, residuals are small. Great variability gives large residuals.

Now we need to compress all the residuals into one convenient number to represent variability. First of all, residuals are positive and negative, so there is no point in averaging them. The mean residual is always zero; see equation (4-2). Instead, we want to get rid of the negative signs in residuals before averaging. One way of doing this is by squaring each residual, $(x_i - M)^2$. Then, we can find the *mean squared residual* and, in order to get back to the original units of measure, take a square root of it:

$$SD_n = \sqrt{\frac{1}{n} \sum_{i=1}^{n} (x_i - M)^2}, \tag{4-7}$$

This is *a* standard deviation, but not exactly as you might know it. The most commonly used formula is as follows:

$$SD_{n-1} = \sqrt{\frac{1}{n-1} \sum_{i=1}^{n} (x_i - M)^2}. \tag{4-8}$$

This is the sample standard deviation you will find in most textbooks. Hence, I am going to use it throughout this book. In this section, I'm going to call it SD_{n-1}, in order to distinguish it from other standard deviation estimators. Later in the book, for the sake of simplicity, I will drop the '$n - 1$' subscript and call it SD.

The only difference between SD_n and SD_{n-1} is $n - 1$ in the denominator in the latter formula, which is called *Bessel's correction*. One of the reasons for using $n - 1$ instead of n is that there are $n - 1$ degrees of freedom in calculating a sample standard deviation. The other reason is more important, but a bit more complicated. I will try to explain this next.

Degrees of freedom in standard deviation are explained in Section 4.8.

Unbiased estimator of standard deviation

Let us forget about standard deviation for a moment and consider the sample *variance* instead. Variance is standard deviation squared. Just like in standard deviation, you can use two estimators of variance, SD_{n-1}^2 or SD_n^2. It turns out that SD_n^2 is biased, whereas SD_{n-1}^2 is unbiased, because including the '$n - 1$' term removes the bias.

A variance estimator with the '$n - 1$' correction is unbiased.

How can we verify this? Let us do another thought experiment. Consider again a population of mice with mean body weight of $\mu = 20$ g and standard deviation of $\sigma = 5$ g. The population variance is, obviously, $\sigma^2 = 25$ g^2. Take a lot (let's say a million, why not?) of small samples ($n = 5$) from this population. For each sample, let's calculate the two estimators of the variance. Having collected a million SD_{n-1}^2 and SD_n^2 values (which I have simulated in a computer), we can plot their distributions, find mean values and compare to the true variance, σ^2. This is illustrated in Figure 4-5 and shows that the mean SD_{n-1}^2 is better in representing population variance than the mean SD_n^2. These sorts of experiments are almost impossible in practice, but an analytical calculation can show the same result. A simple algebraic derivation (e.g., Brandt 1999) can show that the mean uncorrected sample variance is

$$\langle SD_n^2 \rangle = \frac{n-1}{n}\sigma^2. \qquad (4\text{-}9)$$

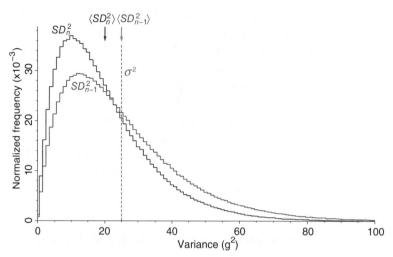

Figure 4-5. Results from a computer simulation to illustrate the differences between two variance estimators, biased SD_n^2 and unbiased SD_{n-1}^2. One million samples of five data points were drawn from a Gaussian population, $\mathcal{N}$ (20, 5). For each sample, two variance estimators were calculated. The graph shows distributions of both estimators; arrows indicate the mean of each distribution; the dashed vertical line shows the true population variance, $\sigma^2 = 25$ g². This simulation demonstrates that $\langle SD_{n-1}^2 \rangle = \sigma^2$ and $\langle SD_n^2 \rangle = \frac{4}{5}\sigma^2$ for the sample size of five.

Hence, on average, SD_n^2 will *underestimate* the true σ^2 by a factor $(n-1)/n$. Luckily for us, the '$n-1$' correction cures the underestimation problem,

$$\langle SD_{n-1}^2 \rangle = \sigma^2. \tag{4-10}$$

This shows that the '$n-1$' corrected variance is an unbiased estimator. Regrettably, not everything in life is so beautiful. Bessel's correction works nicely for variances but not, alas, for standard deviations. This is because averaging and squaring cannot be swapped. The square of the mean is always less than (or equal to, if all numbers are the same – but this is very unlikely in our case) the mean of squares[2]. In particular,

$$\langle SD_{n-1} \rangle^2 < \langle SD_{n-1}^2 \rangle. \tag{4-11}$$

---

[2]This is called Jensen's inequality, and its proof can be found in mathematical textbooks; it is easy to see how it works in a simple example of $x_1 = 3$ and $x_2 = 5$: $\langle x \rangle^2 = \left[\frac{1}{2}(3+5)\right]^2 = 4^2 = 16$, but $\langle x^2 \rangle = \frac{1}{2}(9+25) = 17$.

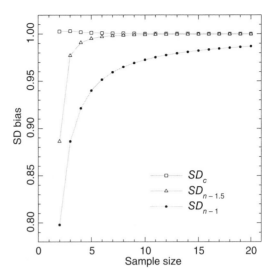

Figure 4-6. Bias of three standard deviation estimators as a function of the sample size. The graph shows how much the population standard deviation is underestimated by each estimator. SD_{n-1} is the default, commonly used estimator, and the other two incorporate simple corrections shown in equations (4-12) and (4-13).

Taking the root of both sides of equation (4-11) and taking into account equation (4-10), we find that $\langle SD_{n-1} \rangle < \sigma$, and, despite all of our efforts, we still have a problem:

> Even the '$n - 1$' corrected standard deviation estimator is biased and underestimates the population standard deviation.

The bias is particularly large for very small samples. For $n = 2$, 3 and 4, the population σ is on average underestimated by 20, 11 and 8% (see Figure 4-6). This is really a shame!

But where is a will, there is a way. You can correct SD_{n-1} (again!) and remove the bias altogether. All you need to do is to multiply SD_{n-1} by a correcting factor. The exact form of this factor is rather complicated, but there is a simple and useful approximation that works well in practice (Gurland and Tripathi 1971):

$$SD_c = \left[1 + \frac{1}{4(n-1)} \right] SD_{n-1}. \tag{4-12}$$

This corrected estimator of standard deviation is approximately unbiased, with accuracy better than 0.3%. Another possible trick, not as effective as SD_c, but exceptionally simple, is to replace $n - 1$ in equation (4-8) with $n - 1.5$. The resulting estimator,

$$SD_{n-1.5} = \sqrt{\frac{1}{n - 1.5} \sum_{i=1}^{n} (x_i - M)^2},$$
(4-13)

is also approximately unbiased for $n > 2$. The bias for $n = 3$ and 4 is 2.3 and 0.9%, respectively (Figure 4-6).

The usage of t statistic to find confidence intervals of the mean is discussed in Section 5.4. The standard deviation estimator is extensively used in all kinds of experimental sciences. It is surprising that the knowledge of SD_{n-1} being biased for small samples is less than universal outside the community of mathematicians and statisticians. On the other hand, standard deviation is typically used to either find a confidence interval or perform a t-test to compare two samples. In both cases, a t statistic is used, which employs the 'default' SD_{n-1} estimator and takes sample size into account (i.e. it gives correct results for a small sample size). When you calculate the confidence interval for the mean, you should use the *uncorrected* estimator SD_{n-1}.

Mean deviation

In this chapter, I showed how to estimate variability in the sample by a standard deviation. The idea was simple (see Figure 4-4): find the deviations from the mean, get rid of negative signs by squaring them and find their mean (or Bessel-corrected mean). But there is another way of getting rid of negative signs, the way that looks simpler on paper. Instead of squaring the residuals, we can take the absolute value,

$$MD = \frac{1}{n} \sum_{i=1}^{n} |x_i - M|.$$
(4-14)

This quantity is called the *mean absolute deviation*[3] (hereafter, mean deviation), and it is *an* estimator of a sample's variability. How does the mean deviation compare to the standard deviation? It looks simpler (there is no square), but in reality it is more difficult to handle in algebraic calculations. Absolute values are a bit tricky when it comes to derivatives ($|x|$ is not differentiable at $x = 0$). On

---

[3]Sometimes denoted as MAD.

the other hand, mean deviation doesn't overestimate outliers as the standard deviation does. The contribution of a data point to *MD* is proportional to the deviation of this point from the mean, whereas in *SD* it is proportional to the square of the deviation. That is why in the example from Figure 4-4 the standard deviation, $SD = 4.8$ g, is greater than the mean deviation, $MD = 3.8$ g. In some cases, occasional strong outliers can distort *SD* to the point at which a researcher considers rejecting them, only because they 'don't look right'. This is not a recommended practice. Perhaps using *MD* to describe data dispersion would encourage treating data with more respect.

Then again, standard deviation is very well established in statistics. It is 'built into' the Gaussian distribution, and one or three sigma probabilities are used widely. It is used to find confidence intervals for the mean, a very important error measure. There are advocates of using the mean deviation in everyday laboratory practice (e.g. Gorard 2005), but the standard deviation, despite its drawbacks, is doing very well and won't disappear from statistics books overnight.

Pearson's correlation coefficient

Correlation is a statistical relationship between two random variables. In practice, we measure the correlation between two sets of data, where data points correspond to each other, so that you can arrange them in pairs. For example, if you have a group of genes of interest, you can measure their expression level under two conditions – let's call them X and Y. For gene i, you will have a pair of expression levels, (x_i, y_i). If genes behave similarly under the two conditions, you would expect the two sets of measurements to be correlated. Another example could be the relationship between the heights of fathers and their adult sons (see Exercise 5.3).

Mathematically speaking, Pearson's correlation coefficient for two sets of (paired) measurements, $x_1, x_2, \ldots, x_n$ and $y_1, y_2, \ldots, y_n$, is defined as

$$ r = \frac{1}{n-1} \sum_{i=1}^{n} \left(\frac{x_i - M_x}{SD_x} \right) \left(\frac{y_i - M_y}{SD_y} \right), \tag{4-15} $$

where M_s and SD_s are the mean and standard deviation of sample s. Quantities in brackets are called *standard scores* or *Z-scores* for each data point and tell us how many standard deviations from the mean the given point is. The correlation coefficient r has a simple

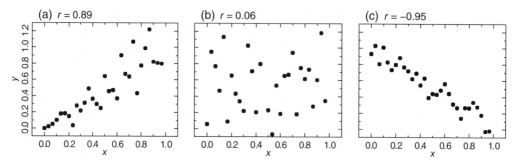

Figure 4-7. Correlation between two variables, x and y. Each panel shows 30 pairs of measurements, (x_i, y_i). Numbers above panels show Pearson's correlation coefficient, r. (a) Positive correlation; (b) no correlation; and (c) strong negative correlation (anticorrelation).

intuitive meaning, which is best explained in a plot. Figure 4-7 shows examples of data with either positive correlation (r close to 1), no correlation (r close to 0) or negative correlation (r close to −1). Two perfectly correlated samples (r = 1) would form a straight line in a plot.

Confidence interval of the correlation coefficient is discussed in Section 5.7. For small data sets, the correlation is usually poorly established. When you have only five pairs of numbers, the correlation coefficient might be quite high, but its uncertainty is going to be huge and the correlation might not be statistically significant. I will discuss this in Chapter 5.

You should exercise caution when interpreting the correlation coefficient. Correlation between two variables does not necessarily imply a direct causal link between them. A famous example comes from Australia, where it has been reported that whenever ice cream sales rise, so do shark attacks[4]. The two variables are correlated, but this doesn't mean that eating ice cream makes you tastier for a shark. A more prosaic explanation is that in hot weather, people both eat ice cream and go to the beach to swim (with the sharks) more frequently.

The estimator described by equation (4-15) is biased, and it underestimates the true correlation coefficient (in absolute values). It is possible to correct for this, but the correction is mathematically complicated, so I will only refer the curious reader to the relevant literature (Olkin and Pratt 1958).

---

[4]This is probably a myth. I couldn't find any actual data confirming this correlation. However, it sounds very plausible and hilarious, so it makes a good example.

Proportion

At first glance, proportion is a very simple thing: a ratio of two numbers. For example, if there are 6 mice out of the initial 10 alive 5 days into the experiment, the surviving proportion is 0.6. If 134 people out of the sample of 998 voted for party X[5], then the proportion is about 0.13. In statistics we use the term *number of successes*, and *proportion* is defined as

$$\hat{p} = \frac{\hat{S}}{n},$$ (4-16)

where $\hat{S}$ is the number of successes and n is the total number of objects in question. Mind you, *success* is just a word; voting for party X might turn out to be a complete failure after all.

As usual, the statistical estimator derived from the sample only roughly approximates the true proportion in the population. The larger the sample, the better the estimator, hence an opinion poll based on three subjects (including one dog) is worthless. In survival experiments the number of animals (e.g. mice) is usually very limited, so knowing the error of proportion is very important.

Confidence interval of a proportion is discussed in Section 5.8.

We can link proportion with the binomial distribution. A sample of size n can be considered as a series of n events with a probability of success p. A random variable S, representing the number of successes, is binomially distributed with the mean np and the standard deviation $\sqrt{np(1-p)}$; see equations (2-9) and (2-10). A proportion of successes can be represented by a scaled random variable, $\varphi = S/n$. If you multiply or divide a random variable by a constant, its mean and standard deviation scale by the same constant. Therefore, the mean of φ is

Binomial distribution is explained in Section 2.7.

$$\mu_\varphi = p,$$ (4-17)

and its standard deviation is

$$\sigma_\varphi = \sqrt{\frac{p(1-p)}{n}}.$$ (4-18)

Proportion is distributed with a scaled binomial law.

[5]The author of this book denies any links to party X, if such a party exists anywhere in the world.

4.5 Standard error

The two most important statistical estimators you can calculate from a sample are the mean and the standard deviation. The mean tells us where the sample is centred; the standard deviation is a measure of the sample spread. This is all beautiful, but we would like to know more. In particular, we would like to know how well the mean is established: we would like to know the error of the mean. Standard deviation is not a good measure of this error, as it doesn't reflect the size of the sample. Surely, with 100 data points we can pinpoint the mean much better than with only 5 data points! The error of the mean should somehow scale with the sample size.

Let us conduct yet another thought experiment. Consider our mouse population again, with body weights distributed normally ($\mu = 20$ g and $\sigma = 5$ g). Let us take a random sample of five mice and calculate the mean, M. Repeat this experiment eight times, taking a different random sample every time. The result is shown in Figure 4-8a. Because the distribution of body weight in the population is rather wide ($\sigma/\mu = 0.25$), measurements from individual samples are also scattered widely. This is reflected by the size of standard deviation from each sample.

I will come back to the sampling distribution in Section 5.1. Now, allow ourselves full play and repeat sampling 100,000 times. I have simulated this in a computer, collected the mean for each sample and plotted the distribution of sample means in Figure 4-8b. This is called the *sampling distribution of the mean*, and it is probably one of the most important probability distributions in this book. It doesn't tell you how individual measurements are distributed; it tells you how the sample mean is distributed, should you perform your entire experiment many times.

Note that the sampling distribution is much narrower than the original Gaussian distribution from which the samples were drawn. The standard deviation of the population is $\sigma = 5$ g, whereas the standard deviation of the sampling distribution from Figure 4-8b is about 2.2 g. The width of the sampling distribution depends on the sample size: the bigger the sample, the narrower the distribution. Figure 4-8c and 4-8d show the same simulated experiment performed for the sample size of 30. Although individual measurements have similar scatter to that in Figure 4-8a, sample *means* are much more concentrated around the true population mean. This is reflected by the much narrower sampling distribution in Figure 4-8d. Remember: all the samples in Figure 4-8 come from the same population, and yet there is a considerable difference between $n = 5$ and 30.

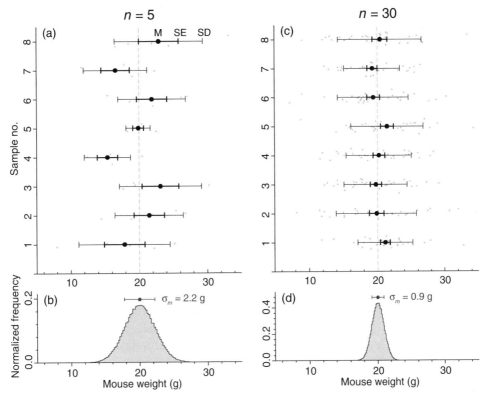

Figure 4-8. Standard error of the mean for the sample sizes of five (left) and 30 (right). Top panels show eight independent random samples. Grey dots represent individual body weights; black circles with longer error bars correspond to sample mean ± standard deviation. The shorter error bars, marked with a thicker line, show the sample standard error. All samples come from a normal population with $\mu = 20$ g and $\sigma = 5$ g. Each of the bottom panels shows distribution of sample means from 100,000 randomly generated samples. This is the sampling distribution of the mean. The dot with error bars above the distribution illustrates the distribution mean, $\mu_m = \mu$, and standard deviation, $\sigma_m = \sigma/\sqrt{n}$, respectively. The sampling distribution of the mean is narrower for a larger sample size, with $\sigma_m = 2.2$ and 0.9 g for $n = 5$ and 30, respectively.

Sample size matters!

You probably see where this is going. The *width* of the sampling distribution can be perhaps used to estimate the uncertainty of the mean. The narrower the sampling distribution, the better we can constrain the mean.

What is this width, then? To find out, let us consider the sample sum, $S = \sum_{i=1}^{n} x_i$, first. A single measurement, x_i, can be treated as a random variable with a certain probability distribution (it doesn't have to be Gaussian), with a mean μ and a standard deviation σ. The sample sum, S, can also be considered as a random variable derived from individual measurements. Because it is a sum of n independent variables, we can use the central limit theorem to predict that it is going to be normally distributed for a large n. The mean of the variable S is simply $n\mu$, just because we added n independent random variables with the mean μ each. We also know (from basic statistics) that variance is additive. If we add n random variables with variance σ^2 each, the resulting variance is $n\sigma^2$. Hence, the standard deviation of the sum is the square root of its variance, $\sqrt{n}\sigma$. We conclude that the sample sum, S, is distributed normally with the mean $n\mu$ and standard deviation $\sqrt{n}\sigma$.

I discussed central limit theorem in Section 2.5.

The sample mean is $M = S/n$ (i.e. the sample sum *scaled* by a factor $1/n$). When we scale a random variable, its mean and standard deviation scale in the same way, so we have to divide them by n. This gives the mean and standard deviation of M of $n\mu/n = \mu$ and $\sqrt{n}\sigma/n = \sigma/\sqrt{n}$, respectively.

To summarize, the sampling distribution of the mean approaches a Gaussian distribution for a large n. Its mean is the same as the population mean, $\mu_m = \mu$, and its standard deviation is

$$\sigma_m = \frac{\sigma}{\sqrt{n}}. \tag{4-19}$$

This is a measure of the uncertainty of the mean we are looking for, but, alas, it is not very useful. We do not know the population standard deviation, σ, so we cannot find σ_m. We can, however, do a trick commonly used in statistics. We can replace the unknown population parameter with the known sample estimator. In this case, we replace σ with SD and find the following estimator:

$$SE = \frac{SD}{\sqrt{n}}. \tag{4-20}$$

Remember the counting error from Figure 3-5?

It is called the *standard error of the mean*. Please note that this is an estimator calculated from just one particular sample. It estimates σ_m, but it is not equal to σ_m. Short, thicker error bars in Figure 4-8a and 4-8c show individual standard errors for each sample. If you look carefully, you will notice that they change from sample to sample (especially for $n = 5$), but they are all comparable with the population σ_m, which is shown in the bottom panels of the figure. This is how estimators work, and you are going to see it many times throughout the book.

Also, the true population mean (usually unknown in real experiments) is not always within the standard error from a sample.
For example, you can see from Figure 4-8a that the true mean is
outside of the *SE* error bars for samples 3, 4 and 7.

*More about probabilistic
interpretation of the
standard error in
Section 5.5.*

There is no guarantee that the true population mean is
within $M \pm SE$.

Standard error, *SE*, is derived from the standard deviation, *SD*.
You can use various estimators of the standard deviation. If you
use the biased SD_{n-1}, you will end up with a biased standard error, which underestimates the error of the mean. You can use the
corrected version, SD_c, as shown in equation (4-12), to obtain an
(approximately) unbiased standard error. This correction should
not be used to find confidence intervals.

Table 4-2 shows the similarities and differences between standard deviation and standard error. Standard error is typically used

Table 4-2.   Comparison of standard deviation and standard   error
estimators. Equations in the top row show the Bessel-corrected
biased estimator SD_{n-1}; however, it can be replaced with the
unbiased SD_c from equation (4-12). The critical value t^* is calculated
from the *t*-distribution and explained in Section 5.4.

Standard deviation	Standard error
$SD = \sqrt{\frac{1}{n-1} \sum_{i=1}^{n} (x_i - M)^2}$	$SE = \frac{SD}{\sqrt{n}}$
Measure of dispertion in the sample	Error of the mean
Gives an estimate of the true population standard deviation, σ	Tells you how accurately you can estimate the mean
About 68% of data points should be within $M \pm SD$, but only if the population is Gaussian.	In about 68% of repeated experiments, the true mean μ is within $M \pm t^* SE$.
Does not depend on sample size	Gets smaller with increasing sample size
To be used if you want to estimate either the variability or dispersion in your data	To be used only to assess the uncertainty of the mean (but it is better to use confidence intervals)

to express uncertainty in the sample mean. This is a standard procedure and there is nothing wrong with it. I would like to point out, however, that the confidence, related to the standard error, varies with the sample size. I strongly recommend using 'proper' confidence intervals to express uncertainty of any estimator, in particular for the sample mean.

Standard error as a confidence interval is discussed in Section 5.5.

> Standard error is a useful measure of uncertainty of the mean; however, it is better to use confidence intervals instead.

Confidence intervals are the topic of Chapter 5.

4.6 Standard error of the weighted mean

Weighted mean is defined by equation (4-3).

Standard error of the mean is expressed by a very simple formula. However, it cannot be applied to the weighted mean, just because it doesn't contain weights. It turns out that there is no simple analytical formula to calculate standard error of the weighted mean. A few approximations have been suggested (Gatz and Smith 1995), and one specific formula (Cochran 1977) seems to work quite well:

$$SE_W^2 = \frac{n}{(n-1)(\sum w_i)^2} \left[\sum (w_i x_i - WM_w)^2 \right.$$
$$\left. - 2M_W \sum (w_i - W)(w_i x_i - WM_W) \right. \tag{4-21}$$
$$\left. + M_W^2 \sum (w_i - W)^2 \right].$$

Here, $W = \frac{1}{n} \sum w_i$ is the mean weight and all sums are over $i = 1, \ldots, n$. This equation might look a bit scary at first glance, but it contains only simple sums and is easy to compute numerically.

Luckily, unless you have data with arbitrary weights, there might be no need for using equation (4-21). When data are weighted by errors, $w_i = 1/SE_i^2$, we can derive error of the weighted mean by error propagation. If we apply the error propagation formula (equation 7-4) to the definition of the error-weighted mean (equation 4-4), we find

See Chapter 7 for error propagation.

$$SE_W = \frac{1}{\sqrt{\sum_{i=1}^{n} \frac{1}{SE_i^2}}}. \tag{4-22}$$

Please note that this formula takes into account only contributions from individual errors. It doesn't care about how data points are distributed. There is a quiet assumption that SE_i are not any arbitrary uncertainties, but real standard errors (hence, the notation SE_i). Each weighted data point can be regarded as a mean with its standard error. Then, as illustrated in Figure 4-8, the spread in the means (its standard deviation) is similar to individual standard errors, $SE_i \approx SD$.

An interesting thing happens when all individual uncertainties (weights) are equal. If our errors are standard errors, we require $SE_i = SD$. In such a case, equation (4-22) transforms into

$$ SE_w = \frac{1}{\sqrt{\sum_{i=1}^{n} \frac{1}{SD^2}}} = \frac{1}{\sqrt{\frac{1}{SD^2} \sum_{i=1}^{n} 1}} = \frac{1}{\sqrt{\frac{n}{SD^2}}} = \frac{SD}{\sqrt{n}}, $$

which is the standard error of the unweighted mean. This is exactly what we expect, because when all errors are equal the weighted mean reduces to the bog-standard arithmetic mean.

Figure 4-3 illustrates the difference between the unweighted and weighted mean and their corresponding errors. The unweighted mean and its error from equation (4-20) are $M = 2.8 \pm 2.2$, whereas the weighted mean and its error from equation (4-22) are $M_w = 1.7 \pm 0.1$. The unweighted error is artificially inflated by three outliers. The contribution from large outlier errors to the weighted error is negligible, because terms $1/SE_i^2$ in equation (4-22) are very small for large SE_i. Hence, this equation effectively 'ignores' huge spurious errors and takes into account only 'proper' errors from other 'well-behaved' data points.

4.7 Error in the error

It is possible to build the sampling distribution of the standard deviation, analogous to the sampling distribution of the mean in Section 4.6. We would take a lot of samples, measure the standard deviation for each of them and build a distribution of all these SDs. The width of this sampling distribution, expressed by its standard deviation[6], is a measure of SD uncertainty. This quantity can be

---

[6]To be more precise: standard deviation of the sampling distribution of the standard deviation (!).

found analytically, and the formula for the relative error in *SD* is (e.g. Brandt 1999):

$$\frac{\Delta SD}{SD} = \frac{1}{\sqrt{2(n-1)}}. \qquad (4\text{-}23)$$

Quoting numbers and errors is discussed in Section 6.4.

This formula also works if you replace *SD* with *SE*. If you wonder why on Earth we would need such a thing, I can reassure you that this is a very useful formula in everyday practice. This can be used as a measure of the *error in the error*. Knowing the level of (relative) uncertainty in your error is crucial when you want to quote it in writing. I will come back to this later in the book.

4.8 Degrees of freedom

This is perhaps a good time to explain the concept of degrees of freedom, which I use a few times in this book. By definition, it is the number of independent pieces of information used to calculate an estimator. Let us consider a sample, $x_1, x_2, \ldots, x_n$. There are n independent values here. If we calculate the sample mean, there are n degrees of freedom in it. However, there are only $n-1$ degrees of freedom when finding the standard deviation (or variance), because we had to calculate the mean in the first place and we lost one independent value by doing this.

Have a look at the following example. Consider a sample consisting of three data points: 2, 3 and 7 (see Table 4-3). The sum of these data points is 12, and the sample mean is $M = 4$. In the next step we calculate residuals, $x_i - M$, which are needed to find the standard deviation. Because of the definition of the mean, the sum of the residuals is always zero (equation 4-2). It constitutes a *constraint*, which removes one degree of freedom. When you know $n-1$ residuals, you can always find the nth residual from this

Table 4-3.   Calculating standard deviation from three data points, 2, 3 and 7. The last row shows the sum of the rows above. Standard deviation is $\sqrt{14/2} \approx 2.65$ with two degrees of freedom.

i	x_i	$x_i - M$	$(x_i - M)^2$
1	2	-2	4
2	3	-1	1
3	7	3	9
Σ	12	0	14

equation. For example, if you knew the first and second residual to be −2 and −1, then the third residual must be 3. There are only $n − 1$ independent pieces of information here. Hence, there are $n − 1$ degrees of freedom for estimating either the standard deviation or variance. Thus, when calculating the mean squared residual (standard deviation), we divide the sum of squared residuals by $n − 1$. This is one of the reasons for the formulation of $SD_{n−1}$, as shown in equation (4-8). For the same reason, we use a t-statistic with $n − 1$ degrees of freedom for finding confidence intervals of the mean (Section 5.4) and $n − 2$ in the definition of the mean squared residual in the linear fit (Section 8.3).

The other reason is that $SD_{n−1}^2$ is an unbiased estimator of variance.

4.9 Exercises

Exercise 4.1
Four independent experiments measured the random movement of DNA in the nucleoplasm. The first experiment collected more data and provided better statistics, and the remaining three experiments were quick follow-ups. They found the diffusion coefficients of 4.3 ± 0.6, 5 ± 2, 8 ± 5 and 6 ± 2, respectively (all in units of $10^{-3}\,\mu m^2\,s^{-1}$). Errors are standard errors. How would you combine these results together, and how would you estimate the combined error?

Exercise 4.2
Using equation (4-15), demonstrate that the correlation coefficient for two identical samples ($x_i = y_i$ for each i) is exactly 1.

Chapter 5

Confidence intervals

Confidence is what you have before you understand the problem.

—*Woody Allen*

Experiments are about measurements; measurements give you numbers. When you weigh a mouse, you get a number. When you measure the distance between two fluorescent dots under a microscope, you get a number. When you assess drug potency, you get a number. The problem is that all these numbers are uncertain. The uncertainty might come from the variability of the subject (each mouse is different) or from measurement errors. Occasionally, measurement errors can be estimated directly (e.g. reading errors), but in many types of experiments there is no way of even guessing the size of error involved. Typically, intrinsic variability and measurement errors are mixed and entangled together, so when we get our final number there is no way of telling how reliable it is.

More about replicates in Section 5.11.

This is why you should perform experiments in replicates. You repeat your measurement under the same conditions to get several answers to the same question. This can be interpreted as taking a *sample* from a (rather abstract) population. Using this terminology, by doing an experiment in replicates you are interested in the population parameter and estimate it from the sample.

> Doing an experiment in *n* replicates corresponds to taking a sample of size *n* from an (unknown) population.

The most commonly estimated parameter is the mean. The standard error can be used to assess the uncertainty of its estimator, the sample mean. Have a look at sample no. 1 in Figure 4-8a. We get five measurements, we find their mean and standard error and we can quote the result as 18 ± 3 g. Quoting the sample mean

Understanding Statistical Error: A Primer for Biologists, First Edition. Marek Gierliński.
© 2016 John Wiley & Sons, Ltd. Published 2016 by John Wiley & Sons, Ltd.

and its standard error is a common practice in experimental sciences. On one hand, it is good because it gives us a standardized estimate of uncertainty of the mean. On the other hand, it is difficult (although possible) to give the standard error some statistical significance.

This is where confidence intervals come in. In this approach, we not only find the best value and its error, but also assign a certain level of confidence to the error. Typically, we say that the measured quantity lies within a range of values with a certain confidence. This range of values is called the *confidence interval*. In this chapter, I will explain how to find confidence intervals for various estimators and what they really mean.

5.1 Sampling distribution

Sampling distribution of the mean is shown in Figure 4-8. I have already introduced the concept of the sampling distribution while discussing the standard error. Now I am going to show how the sampling distribution can help us to understand confidence intervals. I'm afraid I'm going to bore you to death with our little thought experiments, but I really think they are a nice way of clarifying things that otherwise would require even more boring and complicated maths to explain.

Here it is. Out of all the mice in the universe, let us select a random sample of five and weigh each of them. From this sample, we calculate the mean weight, M. This is a statistical estimator of the population mean weight, μ.

Again, imagine we can repeat this experiment 100,000 times and collect 100,000 sample means. The frequency distribution of these sample means is shown in Figure 5-1. Please note that drawing 100,000 samples of five mice is a very different experiment from drawing one sample of 500,000 mice. In the former case we find the *sampling* distribution of mean, and in the latter case we approximate the population distribution. The sampling distribution is *not* an approximation of the population distribution. It represents statistical behaviour of the sample of the given size. The sampling distribution for $n = 5$ is different from the sampling distribution for $n = 30$, as demonstrated in Figure 4-8. This can be done for any statistical estimator, for example the mean, median, standard deviation, proportion or correlation.

The sampling distribution is the distribution of a statistical estimator.

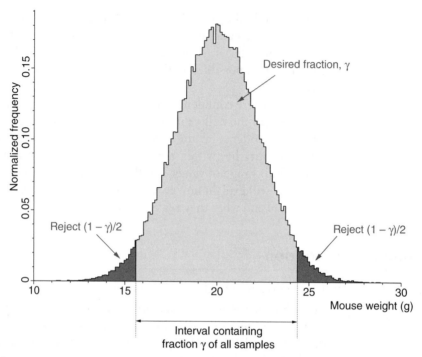

Figure 5-1. Example of a simulated sampling distribution. 100,000 random samples of five points were taken from a normal population of mouse body weight with $\mu = 20$ g and $\sigma = 5$ g. Mean weight was found for each sample. The graph shows the distribution of sample means – the sampling distribution of the mean. The interval indicated by the arrow at the bottom of the figure contains a fraction γ of all sample means, where γ is the requested confidence level.

The simulated sampling distribution of the mean, shown in Figure 5-1, was collected from 100,000 samples randomly drawn from a Gaussian population. As you can see, the sample means are distributed quite widely. To assess the width of the distribution, we can find its standard deviation. From Section 4.5, we already know that the standard deviation of the sampling distribution is called the *standard error*. However, as I will show later in this book, the statistical intuition of the standard error is a bit dubious. We can do better than that and express the width of the sampling distribution from Figure 5-1 in terms of probability, or *confidence*.

Let us choose a particular confidence first, for example, $\gamma = 0.95$ (but see Section 5.3). Sometimes the confidence level is given as a small number $\alpha = 1 - \gamma$, which in this case would be $\alpha = 0.05$. Conventional notation uses the Greek letters gamma and alpha, to denote the 'big' and 'small' confidence levels, respectively. We want to find an interval of mouse body weight, such that the fraction γ (95%) of all sample means falls within this interval. It is very simple, as illustrated in Figure 5-1. All we need is to cut off a total of $1 - \gamma = 5\%$ of outliers on both sides. We probably want a symmetric interval and cut off the same amount of $(1 - \gamma)/2 = 2.5\%$ on each side (dark-shaded regions). This leaves $\gamma = 95\%$ of the samples in the middle. To do this, we can order all 100,000 samples by increasing mean weight and reject the top and bottom 2.5%. The resulting interval in Figure 5-1 is from 15.6 to 24.3 g. It contains 95% of all sample means.

This example shows how to find an interval containing a certain fraction of all possible samples. This is *not* a confidence interval; this example only shows its intuitive meaning. You can think of this as an *ideal* quantity which we will try to estimate.

> The sampling distribution is not used to find confidence intervals in practical applications.

For obvious reasons, you cannot have all the samples shown in Figure 5-1. No funding body will ever let you breed half a million mice for obesity studies. Not to mention animal activists in front of your office. And all the local cats trying to get inside. All you can have is just one sample of n replicates. That's it. Fortunately, there is a way here. As it happens in statistics, instead of repeating the experiment a gazillion times, we can be smarter and predict things theoretically. OK, when I say 'we', I really mean some damn good mathematicians who did it in the past century. For many of the commonly used statistical estimators, it is possible to *estimate* the corresponding sampling distributions analytically and derive confidence intervals from them. In the next few sections, I will show how to do this in practice, but first we need to understand properly what a confidence interval really is.

5.2   Confidence interval: what does it really mean?

Let us consider a 95% CI of the mean. I am choosing this only for illustrative purposes; confidence intervals can be found for (almost) any parameter of interest and for any confidence level. Let us take a (random) sample of 30 mice from the population with the mean body weight of $\mu = 20$ g. Mind you, in real life you don't know the population mean. You are trying to estimate it!

Section 5.4 shows how to find the confidence interval of the mean.

Having all the mice we can weigh them and find the sample mean, M, standard deviation, SD, and the sample size, n (which is usually chosen in advance). From these three numbers, we can

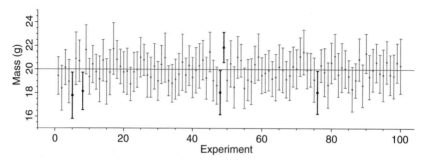

Figure 5-2. The meaning of confidence intervals. An experiment, in which a random sample of 30 mice was weighed, has been repeated 100 times. Each time we found the mean of the sample and the 95% confidence interval of the mean, which are represented by dots and error bars, respectively. We don't know the true mean ($\mu = 20$ g, indicated with the horizontal line in this simulated example), but we know that in 95% of cases it is going to be within our confidence interval. The five cases out of 100 where the true mean is outside the 95% CI are indicated in thicker black lines.

find the 95% CI of the mean. I will explicate later in this chapter how to do this. We still don't know what the true population mean is, but now we have 95% confidence that it is within our interval.

What does this mean exactly? Have a look at Figure 5-2. It shows the body mass experiment with 30 mice repeated 100 times. The error bars encompass 95% CIs for each sample. As you can see, every time the sample mean and the confidence interval are different. Sometimes the true mean is included in the confidence intervals, and sometimes it isn't. The 95% confidence level means that the population mean is included in 95% of the confidence intervals. It is outside the confidence interval in 5% of samples.

> The 95% CI will include the true population parameter in 95% of the repeated experiments.

I would like to warn you about a misconception when it comes to explaining the confidence intervals. The 95% CI *does not mean* that we have 95% probability of finding the population mean within this particular interval. The population mean is *not* a random variable, so we cannot state the probability of finding it within a certain range. The population mean is fixed, even though we don't know it. You have to repeat the experiment many times, finding a *different* confidence interval every time. The true unknown mean will be within 95% of these intervals, but there is no way of telling whether one particular confidence interval contains it or not.

Hence, we cannot make any probability statement from just one sample. When we quote a confidence interval, we implicitly imply the hypothetical statement, 'if we were to repeat our experiment many times …'

> A 95% confidence limit does ***not*** mean you have a 95% probability of finding the population parameter in it.

This is a subtle but important difference.

5.3 Why 95%?

It seems that there is something magical about 95%. Or 5%, depending how you look at it. 95% CIs are commonly used in biology, psychology, social sciences and so on. The limiting p-value of 5% is also popular. The choice of these particular values is completely arbitrary, and probably comes back to the times of the influential statistician Ronald Fisher (1890–1962). He published statistical tables with probability distributions calculated for a few particular probabilities, including 0.05. These tables were used widely by many researchers for many years to come. In his textbook (Fisher 1970) describing the Gaussian distribution and its standard deviation, he said:

> The value for which P = 0.05, or 1 in 20, is 1.96 or nearly 2; it is convenient to take this point as a limit in judging whether a deviation ought to be considered significant or not. Deviations exceeding twice the standard deviation are thus formally regarded as significant. Using this criterion we should be led to follow up a false indication only once in 22 trials, even if the statistics were the only guide available. Small effects will still escape notice if the data are insufficiently numerous to bring them out, but no lowering of the standard of significance would meet this difficulty.

This book, first published in 1925, is regarded to be one of the most influential books of the twentieth century on statistical methods. The 95% CI, or a p-value of 0.05, was, so to speak, imprinted in the minds of many generations of researchers. Nowadays it is commonly used, but, really, you don't have to follow the crowd.

> There is nothing special about the 95% CI.

You can use any reasonable confidence interval as long as you state the corresponding probability. You can equally well use a 90% or 99% interval. If you want to be very stringent (or if you are a physicist, but then, why are you reading this book?), you might use a 'three sigma' limit, which corresponds to the probability of 99.7% (see Table 2-1).

Having said that, I am going to use 95% CIs in this book, just for the sake of consistency. It does not diminish the generality of any of the statements I am going to make. Every 95% confidence given here can be replaced with any other probability, and the same equations can be used.

5.4 Confidence interval of the mean

Population and sample are discussed in Section 4.1. Let us state the problem we are going to solve. There is a population from which we take measurements. For example a population of all mice, each of them having a given body mass, or a population of all possible gene expression levels for a given gene, under certain conditions. The population is characterized by an *unknown* mean, μ, and standard deviation, σ. From this population, we take a sample. It can be a literal sample, like five actual mice, or a set of measurements (replicates). The sample consists of n numbers, from which we can find the sample mean, M, and standard deviation, SD. The question is: how reliable is M? How do we find an interval $[M_1, M_2]$ such that the true mean is included in this interval with a certain confidence, let's say 95%?

Our aim is to find some constraints on the population mean, μ, which is our unknown. The starting point (see Figure 5-3a) is the sampling distribution of the mean, which is the same as in Figure 5-1. It represents all possible samples of size n for the unknown population. We would like to cut out a region containing 95% of samples, as indicated in the plot. Because the width of the sampling distribution is not known (it depends on the population's σ), we can't do it directly. Instead, we can transform this distribution into *Student's t-distribution is* a new distribution of known shape and width, where finding the *described in Section 2.9.* confidence interval is possible. We can do this by calculating the so-called t-statistic,

$$t = \frac{M - \mu}{SD/\sqrt{n}} = \frac{M - \mu}{SE}. \tag{5-1}$$

I have introduced this transformation before. It shows how far our sample mean deviates from the true mean in terms of standard

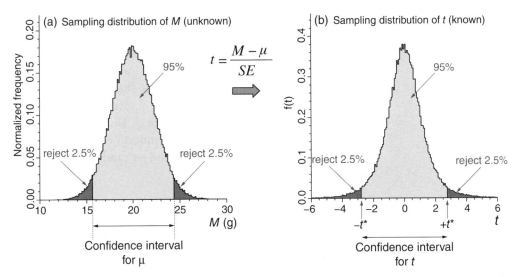

Figure 5-3. Finding the 95% confidence interval of the mean. (a) A simulated sampling distribution (from 100,000 random samples of size $n = 5$) of the mean, M. (b) The corresponding distribution of the statistic t. A simple mathematical transformation converts an unknown distribution of the mean into a known, standardized distribution (Student's t-distribution), from which the required confidence interval can be found.

errors. It is only a mathematical trick, we cannot find t from the sample because we don't know μ. However, we can predict how t behaves statistically. In a thought experiment (where μ is known) we can build its sampling distribution the same way we created the sampling distribution of the mean: by taking lots of samples from the original population and calculating t for each of them. The sampling distribution of the statistic t is shown in Figure 5-3b.

Critical values for t-distribution for selected tail probabilities and degrees of freedom are shown in Table A-1.

And here is the crucial point: it can be shown[1] that the t-statistic follows a Student's t-distribution with $n - 1$ degrees of freedom. The form of this distribution is known analytically, and it is easy[2] to calculate probabilities in any range of t, or vice versa, to find a range of t corresponding to a given probability. You can do this by looking up tabulated values of t-distributions at the end of this book. Alternatively, any self-respecting statistical software can do it. There are also many online t-distribution calculators to do the job.

---

[1] It was first done by William Gosset; see Section 2.9.
[2] Actually, the calculation involves a rather awkward incomplete beta function; when I say 'easy', I mean it can be done quickly and precisely with a computer.

Anyway, it is fairly straightforward to find the so-called *critical value* t^*, corresponding to the required probability, or confidence level, and the number of degrees of freedom, $n - 1$. Figure 5-3c shows a critical value t^*, cutting off 2.5% on each side and leaving 95% probability in the middle (light-shaded area). In terms of probabilities tabulated or computed by software, the right-tail probability is $P(T > t^*) = 0.025$, where T is a random variable with t-distribution. Because the t-distribution is symmetric, 95% of the distribution is contained between $-t^*$ and $+t^*$, so our confidence limit in t-space is

$$-t^* \leq t \leq t^*. \tag{5-2}$$

Now we can reverse our transformation and go back to the mean. Solving equation (5-1) for μ gives us

$$\mu = M - tSE \tag{5-3}$$

If t is between $-t^*$ and $+t^*$, equation (5-3) gives us limits on μ for our particular sample, or, in other words, the confidence limit of the mean:

$$M - t^*SE \leq \mu \leq M + t^*SE. \tag{5-4}$$

Hence, the half of the confidence interval (error) has the size

$$CI = t^*SE. \tag{5-5}$$

This makes things very simple. The confidence interval of the mean is a scaled standard error, and the scaling factor is the critical value from the t-distribution, for a given confidence and $n - 1$ degrees of freedom. Treating half of the confidence interval as error, we can rewrite equation (5-4) as

$$\mu = M \pm t^*SE. \tag{5-6}$$

Selected critical values for the Gaussian distribution are listed in Table 2-1. As I pointed out in Section 2.9, Student's t-distribution approximates Gaussian for larger n. In such case the critical value from the standardized[3] Gaussian distribution can be used instead of t^*. For example, the 95% probability is enclosed within ± 1.96. So, a very crude rule of thumb would be: the 95% CI for the mean is roughly twice the standard error. But it only works roughly and only for

---

[3]Standardized means $\mu = 0$ and $\sigma = 1$.

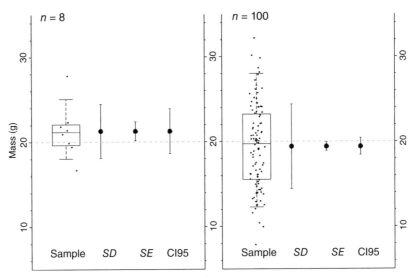

Figure 5-4. Comparison between standard deviation (SD), standard error (SE) and 95% confidence interval of the mean (CI95) for a sample size of 8 (left) and 100 (right). Both samples were drawn from a Gaussian distribution with $\mu = 20$ g and $\sigma = 5$ g.

large samples. I still recommend doing proper calculations using the t-distribution.

A comparison between standard deviation, standard error and 95% CI is shown in Figure 5-4. Both standard error and confidence interval scale with the sample size, and the 95% CI is roughly twice the size of the *SE*.

Example

Consider a sample of seven mice with the following body weights (in grams):

16.8	21.8	29.2	23.3	19.5	18.2	26.3

Let us calculate the 95% CI of the mean. First, we need to find the mean, standard deviation and standard error of our sample:

$$M = 22.16 \text{ g,}$$
$$SD = 4.46 \text{ g,}$$
$$SE = 1.69 \text{ g.}$$

This was very simple. Now, we need to find the critical value from the Student's t-distribution that cuts off $(1 - 0.95)/2 = 0.025$ right-tail probability for $n - 1 = 6$ degrees of freedom. From

Table A-1, we can find that $t^* = 2.447$. From equation (5-5), we get the confidence interval, $CI = 2.447 \times 1.69 \text{ g} = 4.14 \text{ g}$. Finally, we can write

$$\mu = 22 \pm 4 \text{ g,}$$

Rules for quoting numbers and errors are explained in Section 6.4. where the error quoted is the 95% CI. Please note that I have quoted only one significant figure of the error and rounded the mean to the same precision.

Now, let us find a different confidence interval, for example a 99% CI. The critical value for $(1 - 0.99)/2 = 0.005$ and six degrees of freedom is $t^* = 3.797$, and the 99% CI is

$$\mu = 22 \pm 6 \text{ g.}$$

5.5 Standard error versus confidence interval

A confidence interval is a scaled standard error, as shown in equation (5-5), where the scaling factor is the critical value t^*. Let me rewrite this equation explicitly, stating that this scaling factor depends on the sample size, n, and on the required confidence, γ:

$$CI = t^*_{n-1}(\gamma) \times SE. \tag{5-7}$$

I will use this equation to compare the standard error with the confidence interval. We can do this from two points of view.

How many standard errors are in a confidence interval?

For a given value of confidence, the scaling factor gets larger for a smaller n. You can see it from the tabulated values of t^* in Table A-1 in the Appendix. The second column from the right shows the critical values (our scaling factor) for 95% confidence. I visualized these data in Figure 5-5a. It shows how many standard errors are in the 95% CI as a function of n. There is one outstanding point in this plot; for the sample size of $n = 2$, the scaling factor is huge, $t^*_1(0.95) = 12.71$. If you have only two data points in your sample and calculate the standard error, the 95% CI will be almost 13 times larger! This illustrates how little confidence there is in a tiny sample of two points. If you want a sensible result, you definitely need more replicates. For larger samples, the 95% CI corresponds to about two standard errors (the limit for a very large n is about 1.96).

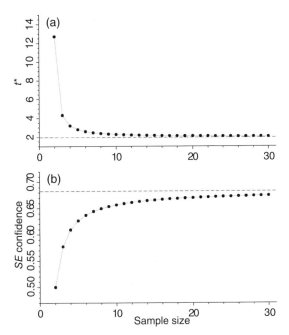

Figure 5-5. Standard error versus confidence interval. (a) The number of standard errors in a 95% confidence interval, $t_{n-1}^*(0.95)$, as a function of the sample size. The dashed line shows the asymptotic $t_\infty^*(0.95) = 1.96$ limit. (b) The confidence corresponding to the standard error for the given sample size. The dashed horizontal line is the asymptotic "one sigma" probability of 68.3%.

What is the confidence of the standard error?

Now let us compare standard errors and confidence intervals from a different point of view. We could ask: what is the confidence corresponding to the standard error? In other words: what confidence γ do we have to request to obtain a confidence interval of the same size as the standard error? From equation (5-7), *CI* equals *SE* when

$$t_{n-1}^*(\gamma) = 1. \tag{5-8}$$

We are looking for a value of γ that gives the scaling factor of 1, as a function of the sample size, n. It is possible to find it from the t-distribution. I plot this relation in Figure 5-5b.

We know already that the standard error is smaller than the 95% CI, so we expect its confidence to be less than 95%. And indeed, it is always smaller than the asymptotic limit of about 68%. You probably recall this number – it corresponds to one sigma in the Gaussian distribution (for a large n, the t-distribution becomes normal). However, for smaller samples this confidence drops substantially.

Statistical confidence is explained in Section 5.2. This is why I advocate against using standard errors in plots, where confidence intervals are feasible. Depending on the sample size, standard errors have different statistical confidence. For example, if you compare two samples of different sizes, repeated experiments will include the true population mean *within the standard error* in 58% and 68% of cases, for $n = 3$ and 50, respectively. This is not a huge difference, but still, it is better to compare things with identical statistical meaning. It doesn't matter what confidence you assume, whether it is 68%, or 95%, or any other, as long as you are consistent.

> It is better to use confidence intervals than to use standard errors.

5.6 Confidence interval of the median

Median was first introduced in Section 4.4. Consider a population of any arbitrary probability distribution. The population median, Θ, divides it into two halves of equal probability,

$$P(X \leq \Theta) = P(X \geq \Theta) = \frac{1}{2}. \tag{5-9}$$

Here, X is a random variable describing our population. This simple fact is always true (by definition!), regardless of the shape of the distribution. Let us draw a sample of n points from the population, $x_1, x_2, \ldots, x_n$, and calculate its median, $\widetilde{M}$. How well does the sample median represent the true population median? Can we find confidence intervals of $\widetilde{M}$?

As before (Section 4.4, 'Median' sub-section), let us sort our sample in ascending order, $x_{(1)} \leq x_{(2)} \leq \ldots \leq x_{(n)}$. The sample median sits in the middle of this sequence. If the sample is truly random and measurements are independent, then the probability that the given point, $x_{(i)}$, is less than the population median, Θ, is exactly $\frac{1}{2}$. This is a simple consequence of the median definition, described by equation (5-9). Obviously, the probability that $x_{(i)} \geq \Theta$ is also $\frac{1}{2}$. By combining these probabilities for all data points, we can find the probability of having k data points to be lower than Θ, and the remaining $n - k$ data points to be higher than Θ. Let's call this probability P_k. You can think of this as the probability of

I used the same argument in Section 5.2. 'finding' Θ between $x_{(k)}$ and $x_{(k+1)}$. This is not exactly true, because the population median is not a random variable and does not have a probability distribution. It is the sample that is random. The actual meaning is as usual: if we draw lots of samples, a fraction P_k of them will have the true median between $x_{(k)}$ and $x_{(k+1)}$.

How can we calculate P_k? Let us use an analogy: we can compare drawing individual data points from the population to tossing a coin. For example, when $x_{(i)} < \Theta$, we call it 'heads'; when $x_{(i)} \geq \Theta$, we call it 'tails'. Drawing n points and checking on which side of the median they lie are the same as tossing the coin n times. The probability of getting k heads (and $n - k$ tails) from n trials

Binomial distribution is described in Section 2.7. can be calculated from the binomial distribution with $p = \frac{1}{2}$, using equation (2-7):

$$P_k = \binom{n}{k} 2^{-n}. \tag{5-10}$$

An example for the sample size of $n = 18$ is shown in Figure 5-6. The sample median is in the middle between the ninth and 10th

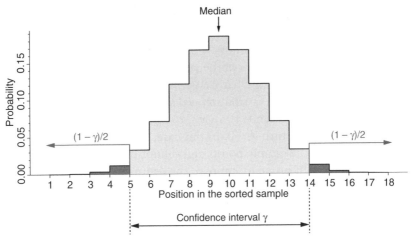

Figure 5-6. An example of the binomial distribution described by equation (5-10). The horizontal axis shows the position (rank) in the sorted sample, that is, 1 corresponds to the first measurement (sample minimum), 2 is the second smallest point and so on. 18 represents the last point in the ordered sample (sample maximum). The sample median (not population median!) lies in the middle, between the ninth and 10th ordered points. The graph shows the probability P_k of 'finding' the population median (see text). The confidence interval, γ, was selected to be as close to 95% as possible. However, due to the discrete nature of the binomial distribution, the actual probability is $\gamma = 96.91\%$. A better approximation of the 95% CI can be found by interpolation (see text).

sorted data points. The probability of 'finding' the population median between these two data points is the same as the probability
of getting nine heads in 18 coin throws, and equals about 18.5%.
There are online binomial distribution calculators that can find
this value for you.

We can now add the binomial probabilities starting from the
middle of the distribution and moving outward, until we get
the desired confidence probability γ, for example 95%. Mathematically speaking, we need to find an index d (between 0 and
$\lceil n/2 \rceil - 1$) such that

$$\gamma_{d+1} \leq \gamma < \gamma_d, \tag{5-11}$$

where γ_d is the probability of 'finding' θ between $x_{(d)}$ and $x_{(n-d)}$:

$$\gamma_d = P(d \leq W \leq n - d). \tag{5-12}$$

Here, W is a binomial random variable for n events and a probability of success of 0.5. The probability γ_d is marked by the
light-shaded area in Figure 5-6. Again, there are online calculators and computer software that can find the cumulative probability from a binomial distribution. In our example, $d = 5$, as
$\gamma_{d+1} = 0.904$ and $\gamma_d = 0.969$. Please note that this gives us only an
approximation of the required confidence interval, and instead of
the 95% CI what we get here is a 96.9% CI. This is because the
binomial distribution is discrete. Our approximate confidence interval extends from $x_{(d)}$ to $x_{(n-d+1)}$. Note that this procedure can
also return $d = 0$. In this case, the confidence interval includes all
of the sample points and cannot be constrained.

In order to improve the estimate, we can interpolate between
$x_{(d)}$ and $x_{(d+1)}$ at one end and between $x_{(n-d)}$ and $x_{(n-d+1)}$ at the other
end of the interval. Having found d, we need to compute the in-
terpolating factor (Hettmansperger and Sheather 1986):

$$I = \frac{\gamma_d - \gamma}{\gamma_d - \gamma_{d+1}}.$$

From this, we can find

$$\lambda = \frac{(n - d)I}{d + (n - 2d)I},$$

which is a quantity between 0 and 1, telling us how to mix the interpolated quantities. Finally, the confidence limits for the median are

$$\begin{aligned}
\tilde{M}_L &= (1 - \lambda)\,x_{(d)} + \lambda x_{(d+1)}, \\
\tilde{M}_U &= \lambda x_{(n-d)} + (1 - \lambda)\,x_{(n-d+1)}.
\end{aligned} \qquad (5\text{-}13)$$

Simple approximation

There is a convenient approximation for quick and easy estimation of the confidence interval of the median, which does not require searching in the binomial distribution (Olive 2005). The sorted sample is $x_{(1)} \le x_{(2)} \le \ldots \le x_{(n)}$. Let $\lfloor x \rfloor$ denote the 'floor' of x, which is the largest integer that is either smaller than or equal to x, and let $\lceil x \rceil$ denote the 'ceiling' of x, which is the smallest integer larger than or equal to x. For example, $\lfloor 3.2 \rfloor = 3$, $\lceil 3.2 \rceil = 4$ and $\lfloor 3 \rfloor = 3 = \lceil 3 \rceil$. Let us define indices

$$L = \left\lfloor \frac{n}{2} \right\rfloor - \left\lceil \sqrt{\frac{n}{4}} \right\rceil, \qquad (5\text{-}14)$$

$$U = n - L.$$

Then, the standard error of the median is

$$\widetilde{SE} = \frac{x_{(U)} - x_{(L+1)}}{2}, \qquad (5\text{-}15)$$

and the confidence interval looks familiar:

$$\begin{aligned}
\tilde{M}_L &= \tilde{M} - t^*\widetilde{SE} \\
\tilde{M}_U &= \tilde{M} + t^*\widetilde{SE},
\end{aligned} \qquad (5\text{-}16)$$

where t^* is the critical value from a Student's t-distribution with $U - L - 1$ degrees of freedom, corresponding to the required confidence level.

Example

Let me present a worked example, which will demonstrate the confidence interval of the median step by step. Imagine a sample of nine mice with the following body weights (in grams):

14.9	22.0	15.0	17.9	21.4	20.6	21.4	24.8	18.0

The sorted sample is:

i	1	2	3	4	5	6	7	8	9
$x_{(i)}$	14.9	15.0	17.9	18.0	20.6	21.4	21.4	22.0	24.8

The sample median is the fifth sorted point, $\widetilde{M} = x_{(5)} = 20.6$ g. From the binomial distribution and equation (5-12), we can find $\gamma_1 = 0.996, \gamma_2 = 0.961, \gamma_3 = 0.820$ and $\gamma_4 = 0.492$. Hence, the condition [equation (5-11)] is fulfilled by $d = 2$:

$$0.820 \leq 0.95 < 0.961.$$

Then, the approximate confidence interval is between $x_{(2)} = 15.0$ and $x_{(7)} = 21.4$ g. Now, we improve our result by interpolation. We calculate $I = 0.0778$ and $\lambda = 0.228$. Finally, from equation (5-13) we find the 95% confidence limits of the median [15.7, 21.9] g. For comparison, the mean is $M = 19.6$ g and the 95% CI is [17.0, 22.1] g.

The approximated method yields $L = 2$ and $U = 7$. From this we get the standard error $\widetilde{SE} = 1.75$ g. The critical t^* for probability 0.025 and 4 degrees of freedom is 2.776 (Table A-1). Hence, the approximate 95% CI is [15.7, 25.5] g.

5.7 Confidence interval of the correlation coefficient

I have introduced Pearson's correlation coefficient in Section 4.4.

Pearson's correlation coefficient, r, is commonly used in experimental sciences to express whether two quantities behave either in a similar way (are correlated) or independently (are uncorrelated). However, there are all too many papers quoting only the correlation coefficient (e.g. $r = 0.73$) without estimating its uncertainty. This is probably OK for a large sample with hundreds or thousands of measurements, but claiming that two quantities 'are correlated', based on only five data points, is dubious. We need to find a way of estimating the uncertainty of r, which will depend on the sample size, n.

Confidence interval of the mean is explained in Section 5.4.

In order to achieve our aim, we are going to follow a similar schema as in finding confidence intervals of the mean. We are going to build a sampling distribution of the correlation coefficient, do a clever transformation into a known theoretical distribution, cut off required probabilities and find corresponding limits on r.

Let us assume a population of pairs of numbers, for which the (unknown) correlation coefficient is ρ. We draw a sample of size

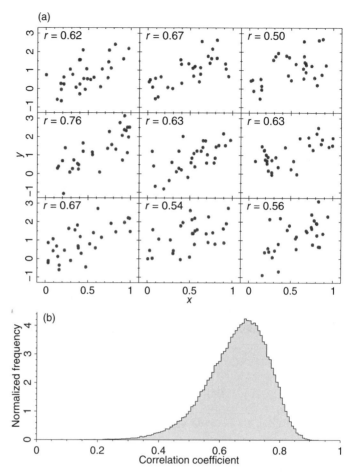

Figure 5-7. (a) Nine examples of samples of 30 (x, y) pairs of points with the correlation coefficient calculated for each of them. The samples are drawn from a population with $\rho = 0.66$. (b) Sampling distribution of the correlation coefficient. 100,000 samples, as above, were drawn from the same population, and the distribution of r is plotted.

n and find this sample's correlation coefficient, r. Let us repeat the procedure many times. Figure 5-7a shows nine samples from the population and their respective correlation coefficients. Figure 5-7b shows the distribution of these coefficients from 100,000 samples. As you can see, this distribution, unlike the sampling distribution of the mean (which is Gaussian), is asymmetric. Fortunately for us, a simple formula called Fisher's transformation,

$$Z' = \frac{1}{2} \ln \frac{1 + r}{1 - r},$$

(5-17)

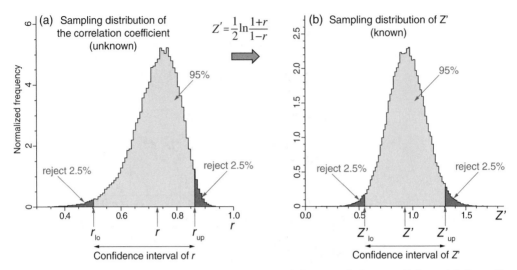

Figure 5-8. Finding the 95% confidence interval of the correlation coefficient. (a) Sampling distribution of the correlation coefficient, *r*. (b) The corresponding sampling distribution of *Z'*. The distribution of *r* is not known. A simple mathematical transformation converts into a known distribution (Gaussian), from which the required confidence interval can be found.

creates a statistic Z', whose sampling distribution is normal, with standard deviation

$$\sigma' = \frac{1}{\sqrt{n-3}}. \tag{5-18}$$

The original and transformed sampling distributions are shown in Figure 5-8. The procedure we apply here is analogous to finding the confidence intervals of the mean. I will show how it works using one practical example. Imagine we have a sample of 30 paired measurements, and the Pearson's correlation coefficient for them is $r = 0.73$. First, using equations (5-17) and (5-18), we find $Z' = 0.929$ and $\sigma' = 0.192$. These are the mean and standard deviation of a Gaussian distribution. Its sampling version is shown in Figure 5-8b. As we know, 95% probability is included within $Z' \pm 1.96\sigma'$ (see Table 2-1), hence the lower and upper limits on Z' are $Z'_{lo} = 0.929 - 1.96 \times 0.192 = 0.553$ and $Z'_{up} = 0.929 + 1.96 \times 0.192 = 1.31$. Now, we reverse the formula (5-17) to find r,

$$r = \frac{e^{2Z'} - 1}{e^{2Z'} + 1}. \tag{5-19}$$

We simply plug Z'_{lo} and Z'_{up} in equation (5-19) to find the corresponding limits on r. When we do this, we find $r_{lo} = 0.50$ and

$r_{up} = 0.86$. Hence, the 95% CI on the correlation coefficient is $[0.50, 0.86]$. This interval is *asymmetric* with respect to $r = 0.73$, which reflects the asymmetry of the sampling distribution shown in Figure 5-8a. We can write it down as an asymmetric plus-minus error,

$$r = 0.73^{+0.13}_{-0.23}.$$

As you can see, the uncertainty is quite substantial even for a reasonably sized sample. The same calculation carried out for the sample size of 6 yields the 95% CI, $[-0.20, 0.97]$. This is a huge error, and you can't say if the two quantities are correlated or not. Hence, if you quote in your paper $r = 0.73$ from a sample of six pairs and claim you have found a correlation, this is, delicately speaking, an unfounded statement. Less delicately, it is utter rubbish.

Significance of correlation

The significance of the correlation coefficient is neither an error nor an uncertainty. However, it is closely related to errors and often used in biological publications, so I thought it deserves a short discussion. Statistical significance is usually based on a null hypothesis. In this case, the null hypothesis is that the sample comes from a population with no correlation at all (i.e. the population coefficient of correlation is $\rho = 0$), and the observed correlation of the sample (e.g. $r = 0.73$, as in Figure 5-9) is purely due to random sampling. It just happened that something emerged out of noise. The probability of this occurring by chance is the statistical significance of the correlation. In Figure 5-9a, the sample is small and this probability is $p = 0.05$. Hence, if we draw many samples of this size from an *uncorrelated* population, one in 20 will show at least this level of correlation. The 95% CI, $[-0.20, 0.97]$, is consistent with zero.

There is a completely different story, however, when the sample size is larger. Figure 5-9b shows a sample with the same correlation coefficient, $r = 0.73$, as in Figure 5-9a, but with 30 data points instead of six. This time, the correlation is highly significant, with the probability of getting it by chance $p = 2 \times 10^{-6}$. This is a respectable value. We can safely claim that these data show a correlation. This is supported by the 95% CI, which is not consistent with zero.

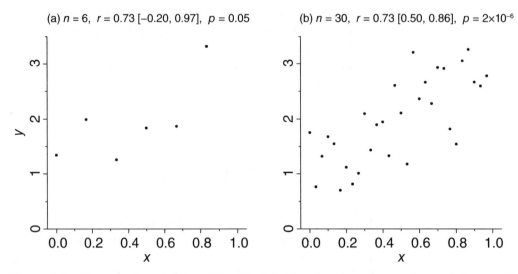

(a) $n = 6$, $r = 0.73$ $[-0.20, 0.97]$, $p = 0.05$ **(b)** $n = 30$, $r = 0.73$ $[0.50, 0.86]$, $p = 2 \times 10^{-6}$

Figure 5-9. Two samples of different sizes but yielding the same Pearson's correlation coefficient of $r = 0.73$. The numbers above the panels show the sample size, n, correlation coefficient, r and its 95% confidence interval and correlation significance.

How do we find the significance of correlation? The method is rather simple. It can be shown that the quantity,

$$t = r\sqrt{\frac{n-2}{1-r^2}}, \tag{5-20}$$

is distributed with a Student's t-distribution with $n-2$ degrees of freedom. The requested p-value of the statistical significance is the one-tail probability from the t-distribution, $P(T \geq t)$, where T is a random variable with t-distribution.

Here is an example. For the samples shown in Figure 5-9a and 5-9b, the t values, found using equation (5-20), are 2-14 and 5-65, respectively. The corresponding numbers of degrees of freedom are 4 and 28. Using either a computer program or an online t-distribution calculator, one can find one-tail probabilities of $p = 0.05$ and 2×10^{-6}, respectively.

We need a one-tail probability because we are asking about the probability of getting a correlation at least as strong as that observed by random sampling. In other words, we want to know the probability of getting a correlation $\geq r$, which is $P(T \geq t)$ in the t-distribution. This is the right-hand tail of the distribution. If the samples were anticorrelated $(r < 0)$, we would be asking for the probability of getting a stronger negative correlation, that is, correlation $\leq r$, which is $P(T \leq t)$, where t is negative. This is the left-hand tail of the distribution. Due to symmetry, $P(T \leq t) = P(T \geq |t|)$, for $t < 0$.

Both here and when finding the confidence interval for the mean (Section 5.4), we use a Student's t-distribution. The way we use it is different, though. When calculating the CI for the mean, we had to find a t-value corresponding to the given tail probability. Here, we do the opposite (i.e. for the given t, we need to find the corresponding tail probability).

5.8 Confidence interval of a proportion

Imagine an experiment in which you have 10 mice. On day 0, you infect them with something nasty and then watch the poor creatures die. Every day, you count survivals and calculate the proportion of the initial 10 that is still alive. The results are shown in Table 5-1. What is the uncertainty of these proportions? Repeated 'measurements' will not help: when you have six mice, you have six mice, regardless of how many times you count them. However, if you repeated the entire experiment once again, you would probably get different counts and different proportions, as a function of time. For example, on day 16 you might still have three mice surviving. Doing survival experiments in many replicates is often impractical, so we need to devise a method of estimating the uncertainty of a proportion from just one sample.

Proportion was introduced in Section 4.4. Consider a large population from which we draw a sample of size n. A proportion $\hat{p} = \hat{S}/n$ of the sample will have the desired property, where $\hat{S}$ is the number of 'successes'. It can be a proportion of voters supporting a specific political party, or

Table 5-1.   Survival experiment. 10 mice are infected with a disease on day 0. The number of surviving mice is recorded every day. The last mouse dies on day 16.

Days	Surviving mice	Proportion
0–3	10	1.0
4–6	8	0.8
7	6	0.6
8–12	2	0.2
13–15	1	0.1
16+	0	0

it can be a proportion of mice alive after a few days of treatment. This is an estimator of the true, unknown proportion in the population.

Sampling distribution is discussed in Section 5.1.

In order to find a confidence interval of the proportion, we have to look at its sampling distribution. As usual, I did a simulated experiment in my computer. I picked 100,000 random samples from a large population with a given proportion of success of $p = 0.134$.

For random sampling see Section 3.4.

The proportion of successful elements varies from sample to sample, due to ... random sampling. From all these samples, I can plot a frequency distribution of sample proportion, $\hat{p}$. I repeated the experiment for three sample sizes of $n = 10$, 50 and 998. The results are shown in Figure 5-10. As in the case of the mean, the width of the sampling distribution (specifically, its standard deviation) is the *standard error* of the proportion.

Proportion follows a scaled binomial distribution, see Section 4.4.

Luckily, we can estimate this standard deviation directly from the sample. We already know that the proportion is binomially distributed and its standard deviation is expressed by equation (4-18). Since we do not know the population proportion, p, we

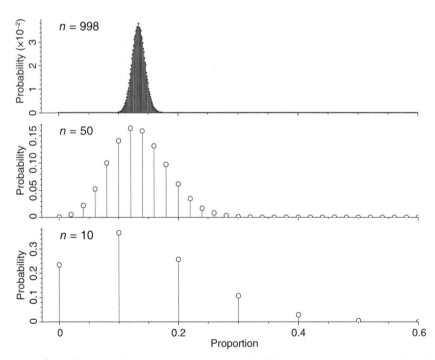

Figure 5-10. Sampling distribution of a proportion. 100,000 samples of size $n = 998$, 50 and 10 were randomly drawn from the population with a true proportion of success of 0.134. Frequency distributions of the sample proportion are shown for each n. With decreasing sample size, the sampling distribution becomes broader and asymmetric.

replace it with its estimator, $\hat{p}$, and get the standard error of the proportion,

$$SE_{\hat{p}} = \sqrt{\frac{\hat{p}(1-\hat{p})}{n}}, \qquad (5\text{-}21)$$

For example, if 134 people out of a sample of 998 support party X, then the proportion is $\hat{p} = 0.134$ and its standard error is $SE_{\hat{p}} = 0.011$. This error scales with $1/\sqrt{n}$ and gets bigger for smaller samples. A similar proportion of 7 out of 50 people ($\hat{p} = 0.14$) would give the standard error of 0.05. For a fixed sample size, $SE_{\hat{p}}$ is largest when the proportion is $\hat{p} = 0.5$.

Now we want to convert the standard error into a more useful confidence interval. The sampling distribution is known (scaled binomial), and for a large n it approximates a Gaussian distribution. Hence, in order to find a 95% CI, you need to multiply the standard error by $Z = 1.96$. For the voters example of $SE_{\hat{p}} = 0.011$, we find $Z \times SE_{\hat{p}} = 0.021$, and the 95% CI is $[0.11, 0.16]$. This is often called the *margin of error* in surveys and opinion polls. You might read in a newspaper, '13.4% of people would vote X; the margin of error is 2%'. The problem is that newspapers will never tell you what confidence interval the margin of error corresponds to. You can only guess if it is 90 or 95%. It can also be the maximum CI, found for $\hat{p} = 0.5$ (in order to have one conservative error estimate for all quoted proportions).

> The 'margin of error' is usually the maximum 95% CI of the proportion. But you never know.

This simple approach does not work well for small samples. As you can see from Figure 5-10, when n gets smaller, the sampling distribution becomes not only broader but also more skewed. Clearly, it is no longer Gaussian, and we cannot simply multiply the standard error by a Z factor. A transformation into a known distribution is not easy here, so a few approximate methods have been developed, to take into account skewness of the sampling distribution and estimate the confidence interval. One of the simplest and reasonably accurate methods is the so-called adjusted Wald method (Agresti and Coull 1998).

Consider a sample of size n, where $\hat{S}$ (number of successes) elements have the desired property. The proportion is $\hat{p} = \hat{S}/n$. First, we need to decide about the confidence level and find the corresponding Gaussian score, Z. For example, for the 95% CI, it is Z = 1.96. The *adjusted proportion* has $Z^2/2$ and Z^2 added to the numerator and denominator, respectively:

Selected Gaussian probabilities are listed in Table 2-1.

$$p' = \frac{\hat{S} + \frac{Z^2}{2}}{n + Z^2}. \tag{5-22}$$

Then, the half-size of the confidence interval (margin of error) is

$$W = Z\sqrt{\frac{p'(1-p')}{n + Z^2}}, \tag{5-23}$$

and the confidence interval extends from $p' - W$ to $p' + W$. If you look carefully at equations (5-22) and (5-23), you will notice their similarity to the standard error [equation (5-21)]. The margin of error is Z multiplied by the adjusted standard error, where modifications replace $\hat{S}$ with $\hat{S} + Z^2/2$ and n with $n + Z^2$. These adjustments bring the estimated confidence intervals close to true values. Note that for the 95% CI. $Z^2 \approx 4$, so when the sample is large (e.g. hundreds of points), and the proportion is not too close to zero, these two adjustments become negligible. Since the sampling distribution is then roughly Gaussian (Figure 5-10), you can use $W = 1.96 \times SE_{\hat{p}}$ as the half-size confidence interval. Note that this confidence interval is asymmetric with respect to the original, uncorrected proportion $\hat{p}$.

Let us go back to our survival experiment. Data from Table 5-1 are shown as a black thick line in Figure 5-11. Lightly shaded large boxes show the 95% CIs of the proportion, calculated using the adjusted Wald method. These errors are quite substantial. For example, when you have eight mice left, the confidence interval of proportion is [0.49, 0.95]. The observed proportion is only an estimator of the true, unknown value. It might be 0.8, it might be 0.6, or it might be 0.94. You don't know. The only thing you know is that in repeated experiments, the true value would be within the calculated confidence interval 95% of the time. There is also a rather large error on the proportion of zero, [0, 0.32][4].

---

[4]The adjusted Wald method cuts off 2.5% of the probability distribution on each side, leaving the 95% CI. When the proportion is 0, the lower limit is fixed at 0, so after cutting off the top 2.5% there is 97.5% left. Hence, the quoted confidence interval is in fact a 97.5% CI, not a 95%

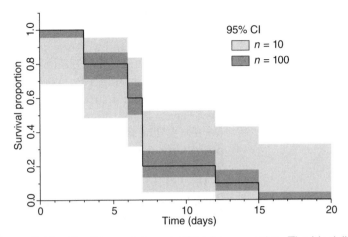

Figure 5-11. Confidence intervals (CIs) for proportion. The black line shows the proportion of mice surviving in an experiment with the initial number of mice, $n = 10$. The lightly coloured boxes show the 95% CIs for this proportion. Darker, smaller boxes show the 95% CIs calculated for the same proportion, but with $n = 100$.

Frighteningly, it seems you have up to three zombie mice at the end of the experiment! You can improve your precision by increasing the sample size (if it is possible to do so). Darker, smaller boxes in Figure 5-11 illustrate the 95% CI for the sample size of $n = 100$.

5.9 Confidence interval for count data

Poisson distribution is discussed in Section 2.7. A Poisson, or count, distribution describes random, independent events that can be counted (e.g. a number of colonies on a Petri dish, or a number of patient deaths over a certain period of time). As I explained before, the standard deviation of a Poisson random variable with the mean number of counts μ is $\sigma = \sqrt{\mu}$. This, however, is a crude estimate of count uncertainty. We can do better than that.

Finding confidence intervals usually involves the sampling distribution. If the sampling distribution is awkward, then we can

---

CI. You can either quote this interval with the appropriate note, or re-calculate it by cutting off the top 5% (i.e. picking $Z = 1.64$) and risking inconsistency with other intervals. The same problem exists for the proportion of 1.

transform it to a familiar form and do our probability calculations there. The sampling distribution of the count number is a Poisson distribution, and we know its analytical form. If we measured the number of bacterial colonies under identical conditions many times, we could build a Poisson distribution out of these counts. However, cutting a specific tail off a discrete distribution is not easy. Have a look at Figure 2-1a: it shows the Poisson distribution for $\mu = 4$. The first bar ($X = 0$) contains about 1.8% of the total probability, whereas the first two bars ($X \leq 1$), encompass about 9.2%. How on earth are we going to cut exactly 2.5% off?

We can do a little trick. We can shift the entire distribution to the left and to the right, by adjusting its mean, until the tail cut off by our measured count number gives us exactly the required probability. I realize that what I just said does not make any sense, so let me explain this using a pretty picture.

A few examples of Poisson distribution are shown in Figure 2-7c. Consider a measured count of $k = 5$, for example five bacterial colonies. Now, draw a Poisson distribution for a given *mean* number of counts, μ. Remember that μ does not have to be an integer number, as it represents the mean. And here is the trick: imagine an interactive picture (which I cannot draw in this rather static book) in which we can 'slide' the distribution to the left and right by increasing and decreasing μ. Let us start with $\mu = 5$ and decrease it until our $k = 5$ cuts off exactly 2.5% on the right-hand side: $P(X \geq 5) = 0.025$. This is shown in Figure 5-12a. Since μ is a continuous variable, we can do it exactly, without interpolation. It turns out that the required mean is $\mu_1 = 1.62$. We have just found the lower confidence limit, $x_L = 1.62$, that corresponds to a 2.5% cutoff. Now we can do the same above $k = 5$. We increase μ until the left-hand tail gives us 2.5%: $P(X \leq 5) = 0.025$. This is achieved for the new mean of $\mu_2 = 11.67$. This is our upper confidence limit.

Sampling distribution is discussed in Section 5.1. This looks nice in a picture, but how do we calculate these values in practice? How do we find the mean that gives the Poisson distribution with the given tail? We would have to calculate the cumulative Poisson distribution and find its reciprocal value. Funnily enough, this can be done exactly, without approximations, and it leads to yet another probability distribution, the χ^2 (chi-square) distribution. I'm not going to discuss the properties of χ^2 distribution; an inquisitive reader can find that in any good statistics book. I can only say that, like the t-distribution, it is characterized by a number of degrees of freedom. Critical values cutting off selected probabilities are tabulated in statistics books and on the web. I will denote as $\chi^2(p, n)$ the value that cuts off the right-tail probability p in χ^2 distribution with n degrees of

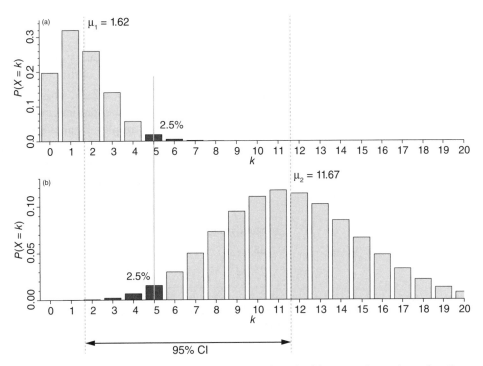

Figure 5-12. Finding confidence intervals for count data. In this example, we have $k = 5$ counts. Both panels show the probability distribution of a random Poisson variable X. First, we move the entire distribution to the left and find a mean, μ_1, such as $P(X \geq 5) = 0.025$. Panel (a) shows a Poisson distribution with the mean $\mu_1 = 1.62$, and the dark-shaded bars show the tail probability of 2.5%. Then we shift the distribution to the right and find a mean, μ_2, such as $P(X \leq 5) = 0.025$, as demonstrated in panel (b). The 95% confidence interval on k is between μ_1 and μ_2.

freedom. Then, the confidence limits on the count number k are (Gehrels 1986)

$$k_L = \frac{1}{2}\chi^2\left(1 - \frac{\alpha}{2}, 2k\right),$$

$$k_U = \frac{1}{2}\chi^2\left(\frac{\alpha}{2}, 2k + 2\right),$$

(5-24)

where $\alpha = 1 - \gamma$ is the confidence level. For 95% CI, we have $\alpha = 0.05$ and the limiting probabilities are 0.975 and 0.025. Please note that χ^2 probabilities are tabulated in various ways. Here I assume that α corresponds to the right-tail probability.

Tabulated confidence limits for count data are shown in Table A-2.   In our example, $k = 5$, we can find tabulated values $\chi^2(0.975, 10) = 3.247$ and $\chi^2(0.025, 12) = 23.337$,   so   our 95% CI is $[1.6, 11.7]$. This can be written as an asymmetric plus-minus error, $k = 5^{+7}_{-3}$. For your convenience, I have created a table of confidence limits for count data, which can be found in the Appendix.

The standard error of the count is discussed in Section 3.5; rules for quoting numbers with errors are laid out in Section 6.4.

A simpler, cruder way of assessing the uncertainty of the count is its standard error, calculated as the square root of the count. From the count number, $k = 5$, we have $SE = \sqrt{5} \approx 2.24$. This error can be written as $k = 5 \pm 2$. As you remember, the standard error of the mean approximates a 68% CI for large samples. By analogy, we can find a 68% CI for count data, using equation (5-24), which, for $k = 5$, gives the interval [2.8, 8.4]. This can be written as $x = 5^{+3}_{-2}$. As you can see, there are many ways the uncertainty of a count can be assessed:

Standard error	5 ± 2
68% CI	5^{+3}_{-2}
95% CI	5^{+7}_{-3}

I cannot stress how important it is to state what type of error you quote. Each of them is different!

Simple approximation

For those who do not wish to use probability tables or who require nonstandard confidence limits, I give a simple analytical approximation that allows quick and fairly accurate estimation of the confidence intervals for count data. Here, k is the number of counts and Z is the Gaussian cutoff for the given confidence limit (see Table 2-1). For example, $Z = 1.96$ for the 95% CI. The approximation is as follows (Gehrels 1986):

$$k_L = k - Z\sqrt{k} + \frac{Z^2 - 1}{3},$$

$$k_U = k + Z\sqrt{k+1} + \frac{Z^2 + 2}{3}.$$

(5-25)

For the 95% CI on $k = 5$, it gives the interval [1.56, 11.75], in a very good agreement with the exact confidence interval [1.62, 11.67].

Errors on count data are not integers

Our best 95% CI for $k = 5$ is [1.6, 11.7]. After rounding, we can write it as [2, 12], which doesn't change the fact that the calculated bounds on the confidence interval are not integers. Once a colleague of mine, who is a physicist like myself, asked me, 'But surely, these errors on counts have to be integers, you cannot have a count of 1.6, you can only have 1 or 2 counts! When you say that $k = 5^{+7}_{-3}$, you mean that k can be between 2 and 12 with a certain degree of confidence. You cannot say that k is between 1.6 and 11.7, right?' It took me a moment to identify a very fundamental flaw in his understanding of errors and confidence intervals. The confidence interval of [1.6, 11.7] is *not* for the measured value of 5. It is for the Poisson mean, μ. And this value can be a real number, for example 3.5, as in our example of radioactive decay shown in Figure 2-7. If we repeat our count measurement many times, and find many (non-integer) 95% CIs, then the true mean will be within calculated confidence intervals in 95% of our measurements.

> Errors on integer counts represent uncertainty in finding the non-integer Poisson mean.

5.10 Bootstrapping

Bootstrapping (resampling) is a computer-based method of assigning an uncertainty to a statistical estimator, in cases where the sampling distribution is either not known or is awkward to calculate. Bootstrapping is rather straightforward, although it requires some programming skills and might be computationally intensive. It is based on a sample of measurements and can be used for any imaginable statistical estimator.

Section 5.4 shows how to calculate confidence intervals of the mean using the t-distribution.

I will demonstrate bootstrapping on a simple example of the arithmetic mean. Since we already know a method of finding confidence intervals of the mean, we can compare it to resampling. Consider a sample of 12 mice, for which we measured body weights. The measurements are (in grams):

19.4, 18.2, 11.5, 17.2, 25.7, 19.2, 21.5, 16.7, 15.6, 27.7, 14.3, 16.3.

The mean, standard deviation and standard error of this sample are $M = 18.61$ g, $SD = 4.59$ g and $SE = 1.32$ g. The critical value from the t-distribution for $p = 0.025$ and 11 degrees of freedom is (Table A-1) $t^* = 2.201$. Hence, the 95% CI can be found as [15.7, 21.5]. This was easy. Now, let us try bootstrapping the original sample.

The method is charmingly simple: we select a random sample from the original sample. The new sample must be the same size as the original (12), but it allows for repeated items. This is called *re-sampling with replacement*. If we were to draw numbered balls from an urn, each ball would be replaced in the urn after calling its number. Hence, the same ball can be called several times. For example, a new sample can look like

15.6, 27.7, 16.3, 15.6, 18.2, 14.3, 16.3, 16.3, 16.7, 17.2, 19.4, 14.3.

In this example, the last measurement of the original sample, 16.3 g, is repeated three times, but the values 18.2, 17.2, 25.7 and 19.2 g are missing. The mean of this sample is 17.33 g, less than the original sample.

Now we repeat this procedure many times, at least a thousand or so. This will create a distribution of bootstrap means similar to that shown in Figure 5-13. Each resampled sample gives us a different 'view' on the original sample. All of them approximate the sampling distribution of the mean. The next step is obvious:

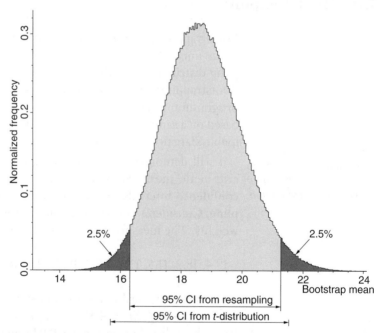

Figure 5-13. Distribution of bootstrap means. The original sample was resampled with replacement 10^6 times. The distribution of the means of all these samples is shown here. Dark-shaded regions show the top and bottom 2.5% of the means. The lightly shaded region in the middle contains 95% of samples and corresponds to the 95% confidence interval (CI). The 95% CI calculated from t-distribution (Section 5.4) is shown for comparison.

we have to cut off the top and bottom 2.5% of all sample means. The easiest way of doing this is by sorting the means in ascending order. The range containing the remaining 95% of samples is our 95% CI. By resampling the original data 10^6 times, I have found the 95% CI = [16.3, 21.3], slightly narrower than the interval found using the t-distribution.

Please note that the bootstrap distribution is not identical to the sampling distribution. The sampling distribution is built by sampling from the whole population – which is impractical in real life. The bootstrap distribution is built from just one sample, which is actually easy to do. It *approximates* the sampling distribution, and this approximation is usually good enough to be applied in practice, so it can be used to estimate the standard error of the mean.

The great advantage of this method is its simplicity and flexibility. Obviously, you don't have to use it to calculate the confidence interval of the mean when the appropriate theory already exists. It can be used for any other statistics, for example the median, where confidence intervals require interpolation. In Section 5.6, I showed an example of a sample of nine measurements for which the 95% CI on the median was found using the binomial distribution with interpolation. The result was [15.7, 21.9] g. Applying the bootstrap to the same sample results in a slightly smaller 95% CI of [17.5, 21.6] g.

Bootstrapping becomes particularly useful when you want to estimate uncertainty of a complex estimator, for example the coefficient of variation[5]. It can also be used for measurements with distributions far from Gaussian or log-normal. Basically, you might use bootstrapping when other methods do not work.

> If all else fails, use the bootstrap.

Bootstrapping has too many applications to mention here; a curious reader can have a look at one of the available textbooks (e.g. Efron and Tibshirani 1993).

5.11 Replicates

Replication is the repetition of an experiment under the same conditions, so that the variability of the measured quantity can

---

[5]The coefficient of variation is the standard deviation divided by the mean.

be estimated. Each of the repetitions is called a *replicate*. Repetition of the same measurement, which does not take inherent biological or technical variability into account, is not replication but *pseudo-replication*. The design of an experiment that avoids pseudo-replication is a big topic (see e.g. Hurlbert 1984).

I mentioned pseudo-replication in the example of counting five mice again and again (Section 2.1) and repeated measurements of the same subject (Section 3.4, Sampling in time sub-section).

I cannot stress enough how important replicates are in experimental biology. In Chapter 3, I tried to demonstrate different types of errors and show how to estimate them. Only a handful of simple measurement errors can be estimated directly either from instrument properties or from theory. In many experiments, errors are too complicated and unpredictable.

> Typically, the only way of estimating measurement errors is to perform the experiment in replicates.

At this point, many biologists ask the inevitable question: how many replicates do I need? Alas, there is no simple answer. The honest answer is: it depends. Depends on your experiment, on what you want to achieve; it also depends on how much funds and time you have. Generally, there are two situations where you need repeated measurements. First, when a particular quantity is measured and you want to find its variability and estimate its error, which is the topic of this book.

The second situation is when you have two (or more) groups of subjects (typically, treatment and control) and you want to compare them. In particular, you want to know whether the measured quantity varies between the groups. This is a part of hypotheses testing and statistical power estimates, and it goes beyond the scope of this book. For more information about sample size in hypothesis testing, see, for example, Van Belle (2008) or Cohen (1988).

Let us briefly discuss the first case. Consider the following example. Figure 5-14 illustrates a (simulated) case of 30 replicates: independent measurements of the same quantity under the same conditions. It shows how the sample mean and its error evolve as we add more replicates. When we have only one replicate (the first point on the left in Figure 5-14), we don't have any statistics: mean doesn't make much sense, and the standard deviation (and error) is undefined. Hence, you don't know the uncertainty of your measurement. You might be bang on the true value, or you might be 50% off. Measurement without error is meaningless.

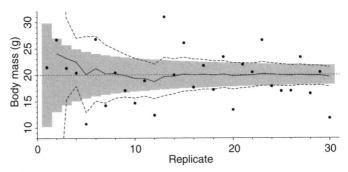

Figure 5-14. A computer simulation illustrating 30 replicates to measure the same quantity under the same conditions. Dots show individual 'measurements', randomly generated from a Gaussian distribution of $\mu = 20$ g and $\sigma = 5$ g. The solid, broken line is the cumulative sample mean. At a given point, n, it shows the mean of measurements from 1 to n. The two dashed lines above and below show the 95% confidence interval for the cumulative mean. The true mean (unknown in real life) is marked with the straight dashed horizontal line. The grey-shaded area shows the 95% variability interval of the cumulative mean: if we repeated the entire procedure many times, 95% of the cumulative means we determine would be within the shaded area (see text).

With two replicates, we can find the mean and get a very rough estimate of the error. The problem is that the sample of only two points is rubbish from a statistical point of view. In our example, the first two measurements are 21.4 and 26.7 g, so their mean is 24.1 g, not very far from the true mean. However, the 95% CI (see Section 5.4) is a whopping ± 33 g! It is so huge that it doesn't fit in Figure 5-14 (dashed broken lines). The standard error, often used to represent the uncertainty, is much smaller, ± 2.6 g, but it corresponds to a very small confidence of only 50%. If we repeat the two-replicate measurement many times, only half of the time the true mean is going to be within the standard error. This is not good. Increasing the number of replicates helps, but not immediately. You need a dozen or so replicates to 'stabilize' the mean and error.

Standard error for a small sample corresponds to low confidence, as shown in Figure 5-5.

The grey-shaded area in Figure 5-14 shows the variability of the cumulative mean calculated as the 95% probability of the sampling distribution. Let me explain how it was calculated. For the given number of replicates, n, I repeated the entire experiment (all n replicates) thousands of times and found the sampling distribution of the mean. Then, I cut 2.5% off each side of the distribution. What is left is a 95% cumulative mean theoretical variability interval. This is not a confidence interval, as explained in Section 5.2, but it would be nice if the actual 95% CI calculated from one sample somehow resembled this 95% variability. And it does, for the large number of replicates! However, when you have only a few replicates, the confidence interval can be quite different than the variability interval (but keep in mind that this is one specific example). Compare the dashed lines and the shaded area in Figure 5-14. A very small sample gives us poor confidence in the mean result.

So, how many replicates do you need? Physicists have a rule of thumb, saying that you need at least 30 replicates to get good statistics. Some biologists I met would happily do an experiment in one replicate. Judging from Figure 5-14, one might say that any sample smaller than 12 is not very good. But, again, it depends on what you need to achieve. Here, I'm going to present just one very simple and intuitive method.

Sample size to find the mean

Sampling distribution of the mean is shown in Figure 5-1. Here is a problem: we want to know the required sample size (number of replicates), to estimate the mean with a given level of accuracy. I will measure accuracy in terms of the confidence interval. Recall the sampling distribution of the mean: it is normally distributed, and the standard deviation of this distribution is $\sigma/\sqrt{n}$. Assume we are interested in a 95% CI error. From Table 2-1, we find it is about 1.96 of the standard deviation, which we can round up to

$$C \approx \frac{2\sigma}{\sqrt{n}}. \tag{5-26}$$

This is not really a confidence interval for the mean, as defined by equation (5-5). Equation (5-26) simply represents the size of the interval containing 95% of the Gaussian distribution. The difference between equations (5-5) and (5-26) reflects the difference between Gaussian and t-distributions, which is not huge unless the sample is very small. In Figure 5-14, the Gaussian 95% is represented by the shaded area, whereas the actual confidence intervals from the sample are shown by dashed lines. We can reverse the relation [equation (5-26)] to find the required sample size,

$$n = \frac{4\sigma^2}{C^2}. \tag{5-27}$$

This is a crude estimation and requires *a priori* knowledge of the population standard deviation. You can have a rough estimate of σ from a pilot study. Coming back to our murine example from Section 2.1, the standard deviation of weight from a sample of mice was 5 g. Assume you want to estimate the weight of a particular strain with accuracy better than ± 2 g (95% CI). From equation (5-27), we find $n = 4 \times 25/4 = 25$. You need at least 25 mice to do this. Please note that n scales with C^{-2}, so to get twice as good accuracy you need a four times larger sample. To estimate the mean weight within ± 1 g, you would need 100 mice.

Once again, equation (5-27) doesn't work for very small samples. If it gives you $n < 5$, you should reconsider your calculations.

5.12   Exercises

Exercise 5.1
The experiment measures expression of a certain gene in wild-type cells (control) and under a certain treatment which is supposed to suppress it (treatment). The control is in 12 replicates; the treatment is in five replicates. The normalized expression levels are as follows:

Control	Treatment
0.787	0.693
0.913	0.657
0.517	0.574
1.295	0.192
0.948	0.739
1.006	
1.208	
1.305	
1.323	
0.789	
0.911	
1.353	

Find the 95% CIs for the mean for each sample. Do they overlap? You can use either the tabulated t-distribution from Table A-1 in the Appendix, any good statistical software, or one of the many available online t-distribution calculators.

Exercise 5.2

A bioluminescent reporter was used to measure transcriptional activity of a gene of interest in a cell culture. The experiment was performed in three biological replicates on day 1, and then repeated in five replicates on day 2. A control was used for normalization. The normalized results are as follows:

Day 1	0.89	0.92	0.89		
Day 2	0.55	0.76	0.61	0.83	0.75

Find the mean and 95% CI for each day. What can you say about the true mean? How else can you use these data?

Exercise 5.3

A small-scale study in a local population tries to answer the question 'Is human height correlated between father and son?' Twelve fathers and 12 adult sons were measured:

Height of father (cm)	Height of son (cm)
172.1	174.8
173.8	172.0
164.5	176.6
181.8	184.3
175.6	175.1
170.2	174.3
175.3	176.5
177.8	173.7
166.4	171.4
166.1	165.8
177.8	181.3
182.1	177.4

Find the correlation coefficient between these two quantities. How strongly are they correlated? What is the uncertainty of the obtained correlation? What's its significance? What is your interpretation of the result?

Exercise 5.4

Consider a large population of Scottish Highland midges (*C. impunctatus*). Through the mystic knowledge of the local clan chiefs (obtained for a bottle of the finest whisky), you know that exactly 13.4% of all midges are completely impervious to the popular repellent DEET. You created three genetically modified strains of midges with the hope of eradicating this resistance. You have obtained a sample from each strain and found the following numbers for DEET resistance:

Strain	Sample size	Resistance count
1	988	115
2	50	3
3	5	2

Calculate resistance proportions and their 95% CIs for each strain. Are they consistent with the mystical true proportion of 13.4%? Is strain 2 promising?

Exercise 5.5
How can you use bootstrapping to estimate the standard error of the mean?

Exercise 5.6
Suppose you have an online tool that calculates the cumulative binomial probability distribution, $P(W < k)$ and $P(W \leq k)$, where k is a given integer number. How would you use it to find γ_d, defined by equation (5-12)? Compute γ_2 for $n = 9$.

Chapter 6
Error bars

Errors using inadequate data are much less than those using no data at all.
—*Charles Babbage*

This chapter is going to be subjective, biased and prejudiced. I will try to force my own personal preferences upon you. I will bicker about incomprehensible plots commonly found in scientific publications. If you agree to endure this, please read this chapter and learn how to improve your plots. If you don't agree, read it anyway.

The topic of this chapter is a bit broader than the title suggests. I will try to elaborate on graphical presentation of data in a more general way. After all, a good plot can relay important information about data and results quickly and efficiently. Some data are almost impossible to present without plots. Throughout this book, I have used plots to illustrate key concepts of statistics. We might recycle an old saying:

One plot is worth a thousand numbers.

Here I will consider only plots presenting numbers. In most cases these will consist of two axes, where one number is plotted against the other. One of the axes might show a categorical variable. For obvious reasons, I will not discuss figures with gels, protein structures, interaction pathways, heat maps and so on. At the end of the chapter, I will discuss how numbers and their errors should be quoted in publications.

6.1 Designing a good plot

Figure 6-1 shows a simple plot you might see in a publication. It contains all the elements of a good plot. It shows a simulated

Understanding Statistical Error: A Primer for Biologists, First Edition. Marek Gierliński.
© 2016 John Wiley & Sons, Ltd. Published 2016 by John Wiley & Sons, Ltd.

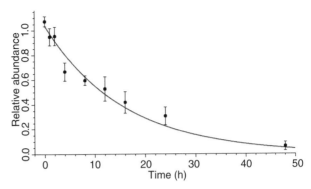

Figure 6-1. A simple plot showing exponential decay of a protein in a simulated experiment. Error bars represent propagated standard errors from individual peptides. Errors in time are considered to be negligible. The curve shows the best-fitting exponential decay model, $y(t) = Ae^{-t/\tau}$, with $A = 1.04 \pm 0.05$ and $\tau = 16 \pm 3$ h (95% confidence intervals).

exponential decay. The horizontal axis represents time, while the vertical axis shows the relative abundance. The data consist of nine measurements, and are presented in the form of a scatter plot, with one black dot corresponding to one measurement. Vertical error bars show uncertainties of measurements. There are no horizontal error bars in the plot. This is because here time is the *explanatory variable* (i.e. the parameter that is set by the experimenter). Obviously, there is a level of uncertainty in this number, in this case resulting from the sample collection protocol, but we consider errors in time to be negligible.

I will explain the terms "explanatory" and "response" variable in Section 8.1.

The measured quantity is the *response variable*. The explanatory variable is usually plotted on the horizontal axis. However, sometimes you might want to plot two response variables against each other, to test how they are correlated. In this case you should plot error bars (if available) in both axes. Generally, you should *always* plot error bars in plots. There are a few exceptions, which I discuss later in this chapter. The curve in Figure 6-1 is an exponential model fitted to the data. The best-fitting model parameters, with their fitting errors, are given in the caption.

Elements of a good plot

As I mentioned before, a good plot enables the reader to see important facts about the data. Because we want to communicate with the reader efficiently, clarity of the plot is vital. There is no point in showing a plot which is either unclear or unreadable.

> When making a plot, always keep clarity of presentation in mind.

The most basic elements of a good plot are:

- axes with scales and labels;
- data represented by symbols, lines, bars or boxes;
- error bars representing uncertainties; and
- model lines, where appropriate.

Each axis should be appropriately labelled. The label should contain the name of the quantity plotted and its units, quoted in brackets. Make sure the name is short, such as 'abundance', 'distance' or 'normalized frequency' and not 'distance between tetO and lacO tandem arrays'. If you have to break the label into several lines, you should probably reconsider the wording. It is better to use a mathematical symbol, such as Θ, and explain its meaning in the figure caption rather than write a long and convoluted description in the axis label. Units should follow conventions used in physics. For example, you should use 's', not 'sec', and 'μm' instead of 'microns'. Obviously, when the quantity is dimensionless (as the abundance in Figure 6-1), units are not necessary.

Make sure that all labels are easy to read. Adjust the font size accordingly, so it is not too small. Don't make it too large. Labels should not dominate the plot.

Symbols representing data should be used with care. See Figure 6-2 for a few examples of bad and good applications. Always keep clarity in mind. Use simple symbols (circles, squares, triangles) and avoid complicated star-like shapes (unless you run out of simpler shapes). When you show two or more data sets in one plot, use symbols easy to distinguish. Symbols should be large enough to recognize, but not too large so they overlap excessively. When you have lots of points (in particular with multiple data sets), you might consider joining them with a line instead of using symbols (Figure 6-2d), but *only* when it improves the overall clarity of the figure.

Colours can be used to differentiate between data categories in the same plot. Selection of appropriate colours is a big topic on its own, and I'm not going to discuss it in this book.

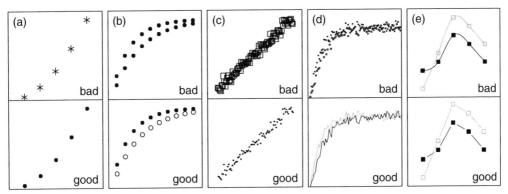

Figure 6-2. A few examples of bad (top panels) and good (bottom panels) application of symbols and lines in plots. These are simplified plots for illustrative purposes only. (a) Use simpler symbols. (b) When presenting multiple data sets, use a different symbol for each set. (c) Scale symbol sizes to avoid overlapping; with a huge number of points, use little dots. (d) Sometimes you might consider joining the data points with a line to make the figure clear. (e) Interpolated lines suggest an underlying model which actually does not exist. You can connect data points with straight lines, as long as it is clear that they are for guidance only. This can be visually highlighted by leaving little gaps between lines and symbols.

Lines in plots

Don't be tempted to use interpolated lines (e.g. splines) to join data points. They might look nice and smooth, but they can be very misleading. The grey interpolated line in Figure 6-2e (top panel) implies a peak (maximum) just right of the middle data point. However, we don't know how the observed system behaves between our data points, unless we take more measurements to fill in these gaps! The problem with interpolated lines is that the reader *might* interpret them as a model, fitted to the data. It is probably better to plot these points without any lines. If you feel that lines are required to guide the eye, use straight-line segments, as shown in the bottom panel. Straight lines do not imply a fitted model, and, hopefully, the reader should understand that they are there for guidance only. You can use an old-fashioned trick and leave small gaps between lines and symbols, as in my figure, to accentuate their guidance-only character.

Some guides to making better plots will tell you that connecting data points with lines is not allowed. I would say: it depends. Generally, yes, you should avoid any unnecessary lines, except for model and trend lines. However, in some cases, particularly when you present many intersecting data sets in one plot, connecting lines might improve plot clarity.

It is OK to join data points with lines, but only when it is necessary for clarity of the plot.

If you decide to use such lines, you should exercise caution. You might create two versions of the plot, with and without lines, and ask a colleague which of them is easier to read. It must be absolutely clear for the reader that these lines are for guidance only and do not represent any model. It might be a good idea to make a note in the figure caption.

In some cases, the dependence between variables shown in the plot (let's call them x and y) can be described by a mathematical model. In the simplest case, where y is (roughly) proportional to x, the relation can be described by a straight line, often called a *trend line*. Linear regression might be used to find the best-fitting line, which can be added to the plot. It shows the essence of the relation between x and y. It also shows the deviation of data points from the theoretical relation. Obviously, the model can be more complicated than the straight line. The exponential decay shown in Figure 6-1 is analytical, but any type of numerical model, for example a Monte Carlo simulation result, can be fitted to data and plotted there.

See Chapter 8 for simple linear regression.

A digression on plot labels

You might have noticed that some plots in this chapter don't have axis labels or even tick marks and numbers. In real research, it would be totally unacceptable. I have done it here only for clarity, as some of the figures would be too cluttered otherwise. I'm not presenting any results here, so actual numbers and labels are not essential. These plots show only one aspect of plot making, so I made them as simple as possible. I still feel somewhat bad about it, hence the disclaimer.

Nevertheless, this doesn't make a valid excuse when presenting real data. Most journal editors would politely ask you to correct plots with missing labels, so it is not easy to publish an unlabelled plot in a reputable journal (although it happens!). Unfortunately, I have seen all too many poor plots in seminar and conference talks, where people show whatever they fancy. Speakers often 'forget' to add axis labels, or labels are small, blurred and unreadable. A poorly rendered picture copied in a hurry from the published paper doesn't help. Such a plot might look perfectly clear to the speaker, who spent days or weeks making it and knows every wiggle by heart. The audience, who are not necessarily familiar with the subject, will only see an obscure collection of dots and hear a vague comment such as 'as you can see, it grows here and drops at the end'. They will have no clue what the plot was about.

Plots should be clearly labelled in talks and presentations!

Logarithmic plots

Log-normal distribution is discussed in Section 2.6. Many quantities in biology are log-normally distributed. They tend to have a huge dynamic range, which is difficult to encompass in a linear scale. It is better to plot such data in a logarithmic scale. The logarithmic scale can either show logarithms of data (e.g. −2, −1, 0, 1, 2) or numbers that are logarithmically distributed (e.g. 0.01, 0.1, 1, 10, 100). Base 2 logarithmic plots are quite common in biology, although I recommend using logarithms to base 10 (see note at the end of Section 2.6).

A logarithmic plot is designed to show data spanning many orders of magnitude in one simple picture. A few examples in Figure 6-3 present a comparison between linear and logarithmic versions of the same plot. When data vary between 20 and 10,000 in some arbitrary units, it is almost impossible to present it in a linear plot (Figure 6-3a). The first three points sit tightly on the horizontal axis and are almost indistinguishable. In contrast to this, the logarithmic plot nicely shows a relation between the data points and wins on clarity of presentation. The vertical axis in this figure has logarithmically distributed numbers.

Figure 6-3b and 6-3c show the same data coming from an RNA-seq[1] experiment. The quantities in the plot are gene expression levels (normalized counts per gene in units of FPKM[2]), with one dot corresponding to one gene. These data are roughly log-normally distributed and cover a whopping six orders of magnitude. The linear plot (Figure 6-3b) shows some correlation between the two replicates, but the bulk of the data are clustered at

---

[1]This is a method of measuring mRNA abundance in order to estimate gene expression levels.
[2]Fragments per kilobase of exon per million fragments mapped.

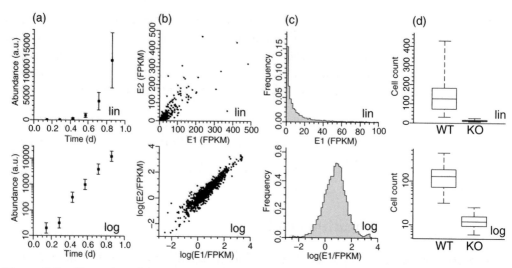

Figure 6-3. Examples of data shown in linear (top panels) and logarithmic (bottom) plots. (a) Simple data spanning three orders of magnitude. (b) Gene expression from two replicates plotted against each other. Note that data points with expression greater than 500 FPKM are not shown in the linear plot. (c) Distribution of expression levels from the same experiment. (d) The two samples represented by box plots are log-normally distributed.

See Figure 2-4 for similar data.

small values and difficult to resolve. Also, the measured expressions span about six orders of magnitude, and in order to show the correlation between points with expression below ~100 FPKM, I had to exclude some of the highly expressed data from the linear plot. You cannot easily fit it all! In the logarithmic version, all data are shown, and we can see a nice linear correlation. Figure 6-3c shows distributions (histograms) of these data. The linear plot is not informative, and it is really difficult to say how the data are distributed by looking at it. The distribution of logarithms tells it all. It is near-Gaussian and reveals a little shoulder at high expression rates.

A very special way of presenting data is a box plot, which I will discuss in detail in Section 6.2. Figure 6-3d shows two data sets, spanning about two orders of magnitude, displayed as box plots. Again, in the linear scale one of the boxes looks tiny and squashed, whereas the logarithmic plot shows both boxes clearly, making their comparison easy.

6.2   Error bars in plots

Traditionally, an error bar is drawn as a line with caps (see Figure 6-4). Both in a plot and in numerical notation, an error can

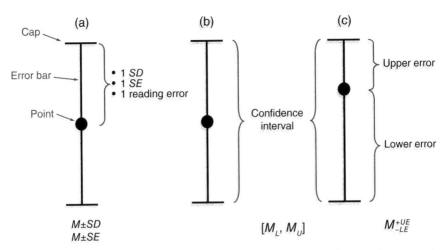

Figure 6-4. A quick guide to plotting error bars. (a) A 'plus-minus' error, for example, 5 ± 3, where error bar extends from 2 to 8. (b) Symmetric confidence interval, which can be written as, for example, [2, 8]. (c) Asymmetric confidence interval, which can be written as, for example, 6^{+2}_{-4}.

be either symmetric or asymmetric. A symmetric error is most common. It is usually denoted as a 'plus-minus' error, for example 5 ± 3. The error quoted (± 3) constitutes a half of the entire error bar, so the bar extends from 2 to 8 in this case. An asymmetric error, such as 6^{+2}_{-4}, has 'plus' and 'minus' bars of different lengths (Figure 6-4c). Confidence intervals for either the correlation coefficient (Section 5.7) or a proportion (Section 5.8) are usually asymmetric.

Various types of errors

See Table 4-2 and Figure 5-4 for SD, SE and CI comparison.

An error bar can represent various things: standard deviation, standard error or a confidence interval (see Table 6-1 for a summary). Occasionally, you can have a Poisson error (which is a standard error) or a measurement error. More complicated uncertainties might result from error propagation (see Chapter 7). Each of these errors is different, so it is absolutely essential to inform the reader which one you used.

Always state what type of uncertainty is represented by your error bars.

Table 6-1.   Types of errors commonly used in graphs. I recommend using confidence intervals in most cases.

Error bar	What it represents	When to use
Standard deviation	Scatter in the sample	Comparing two or more samples, although box plots make a good alternative
Standard error	Error of the mean	The most commonly used error bar, although confidence intervals have better statistical intuition
Confidence interval	Confidence in the result	The best representation of uncertainty; can be used in almost any case

There is a vital distinction between the standard deviation and the standard error. When you want to demonstrate how widely the sample is scattered, you might use the standard deviation. However, box plots (described later in this section) might be better suited for this purpose. If you want to show the uncertainty of the mean, you should use the standard error. But then again, the standard error, although commonly used, does not have a very good intuition in terms of statistical confidence (see Section 5.5).

See Section 8.6 for more details on χ^2 fitting. When you fit a curve to your data points using χ^2 fitting, you need to use 'one sigma' errors from a Gaussian distribution. If each data point represents a sample mean, then you should use the standard error. This is because the standard error represents the (one sigma) standard deviation of the sampling distribution of the mean (see Section 4.5).

In most scenarios, it is better to use confidence intervals to express uncertainty. They show the actual statistical confidence in presented values. Confidence intervals can be calculated for *Chapter 5 shows how to find confidence intervals for various statistical estimators.* most statistical estimators used in everyday lab practice. They also make comparison between two or more samples easier (see Exercise 6.2).

> In most cases, confidence intervals should be used for error bars.

How to draw error bars

Figure 6-5 shows a few simplified examples of properly drawn error bars in various types of plots. Figure 6-5a shows a case where

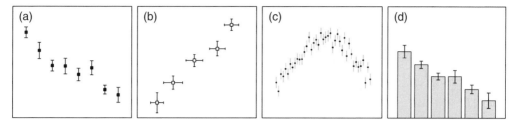

Figure 6-5. Examples of error bars. These are simplified plots for illustrative purposes only. (a) Only the response variable (vertical axis) has error bars. The explanatory variable (horizontal axis) has either negligible errors, or it might be a categorical variable. (b) When both variables have errors, use crosses. (c) In a crowded picture, error bars should be simplified and, perhaps, greyed out. (d) A bar plot with error bars.

See Section 8.1 for explanatory and response variables. the variable on the horizontal axis either has no measurable error (e.g. when it is an explanatory variable) or its error is negligible. In such a case we would plot only vertical error bars, using the elements shown in Figure 6-4: bars and caps. Figure 6-5b shows an example of two response variables plotted against each other. This might be a comparison between two replicates, a treatment versus control, or any two quantities of interest, whether measured or derived. Both variables have errors, so we need to plot vertical and horizontal error bars. Again, we use bars with caps. If there is an underlying model, or any expected relation between the variables, we could add a trend or model line.

Figure 6-5c shows an example of a crowded plot, with many data points. Using the usual error bars with caps (as in Figure 6-5a) would create a horrible mess. Instead, I excluded the caps and plotted error bars in grey, so they don't dominate the picture. Again, clarity is essential.

The next panel shows how error bars should be drawn on top of a bar plot. Typically, an error either with or without a cap is used. No symbol for the data point is necessary, as it is represented by the bar. I will elaborate on bar plots later in this section.

Box plots

Another example of a box plot is shown in Figure 1-1. Figure 6-6 illustrates box plots, sometimes referred to as box-and-whisker plots. They are usually used to compare different categories of data (e.g. various treatments), hence the horizontal axis is typically a categorical axis, as in Figure 6-6a. However, with a reasonably small number of data groups, you can make box plots against a discrete numerical variable (Figure 6-6b).

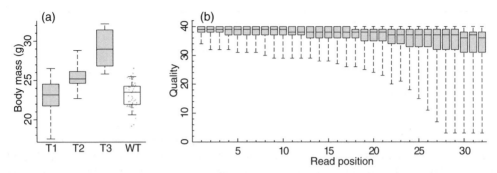

Figure 6-6. Examples of box plots. The line in the middle is the median (50th percentile), the top and bottom of each box show the 25th and 75th percentiles, while the whiskers extend from the 5th to 95th percentiles. (a) Each box corresponds to a different data category. The wild-type (WT) box includes a 'cloud' of data points. (b) Box plots versus a numerical axis, representing DNA short read quality distribution versus the nucleotide position in the read.

Individual boxes show data percentiles: median in the middle, the box itself usually extends from the 25th to the 75th percentiles (i.e. it contains the central 50% of data). The whiskers encompass 90% of data, extending from the 5th to the 95th percentiles. Although the box almost universally represents the 25th to the 75th percentile in most publications, there is no consensus on how to draw the whiskers. Sometimes they represent the minimum and maximum of the sample, sometimes the 2nd or 9th (and 98th or 91st) percentiles, or yet more complicated statistical constructs. Make sure to describe your box plots in figure captions.

The great advantage of a box plot is that it is non-parametric (i.e. it does not require deriving any particular estimators). It shows pure, model-independent data in a simplified form. It can quickly expose asymmetry (skewness) in the sample's distribution. In particular, box plots can reveal differences between samples (e.g. control and treatment). A histogram shows data distribution more precisely, but usually takes more space in a plot, which might be an issue when several data sets are to be compared.

One of the nice additions to a box plot could be a cloud of points. This is shown in the last box (labelled 'WT') in Figure 6-6a. Each data point in the sample is represented by one dot. To maintain clarity, I dispersed the points horizontally, using a random number generator. Otherwise, all these points would collapse into one jumbled line. Some popular graphical packages create box plots with outliers (points outside the whiskers) plotted as dots in one line. This is fine when your sample is not very large. However, since 10% of data are typically outside the whiskers, it can produce quite a few points on top of each other. A slight horizontal scatter should alleviate the trouble.

Bar plots

This is the bit where I'm going to get very personal and prejudiced. For some reason, bar plots are very popular among biologists and used frequently. In my opinion, they are often misused and abused. Let me explain why.

A bar plot is a plot consisting of bars, as in Figure 6-5d. Each bar is a rectangle extending from zero (on the vertical axis) to the value it represents. As bars are usually shaded, what you perceive when looking at a bar plot is the area of each bar. If you plot two shaded bars (rectangles) next to each other, the visually perceived area of these two bars equals the sum of the areas of each bar. In other words, the area is additive. Hence, to make any sense, the quantities presented in the bar plot should be additive too.

> Bar plots should only be used to present additive quantities: counts, proportions and probabilities.

They are perfect for making histograms and probability distributions, as shown in many figures in this book. The area represented by the bars makes perfect sense in such plots. Have a look at Figure 2-1a. Each bar shows a probability $P(X = k)$. These probabilities are additive, and bar areas are additive. The probability of X being between 5 and 7 is the sum of probabilities for $X = 5, 6$ and 7. In the plot, it is represented by the total area of the three dark-shaded bars.

However, it doesn't make any sense to make a bar plot of speed, temperature or distance, as is frequently done in biological publications. These quantities are not additive, so a bar plot suggesting otherwise would be visually misleading.

The entire idea of a bar plot is that the area of each bar is proportional to the value it represents (see Figure 6-7a). Therefore, the bottom (or top, when presenting negative values) end of each bar has to be at zero. If this obvious rule is not followed, the bar plot is simply incorrect, as shown in Figure 6-7b.

> Each bar has to start at zero.

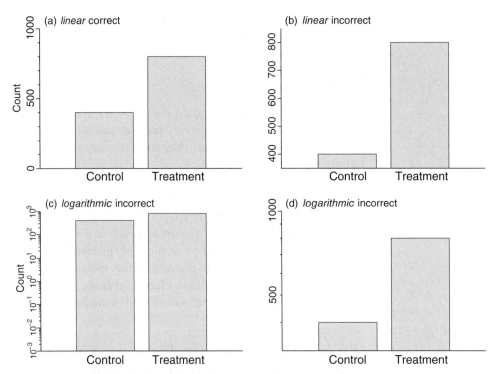

Figure 6-7. Bar plots should always start at zero in the vertical axis. All four panels show the same two data points of 400 and 800 for control and treatment, respectively. Panel (a) correctly shows that the treatment is twice the size of the control. Panel (b) has the vertical axis zoomed in, which alters the perceived difference between the two conditions; in this case, the visible area of the 'treatment' bar is nine times larger than the area of the 'control' bar, which is misleading. The same problem occurs when using the logarithmic scale in panels (c) and (d). Here the lower limit in the vertical axis is completely arbitrary (it cannot be zero!), making the visible bar areas arbitrary as well.

This also means that a bar plot done in logarithmic scale is a complete no-no. Things that are additive in linear scale (Figure 6-7a) are no longer additive when presented in logarithmic scale (the sum of logarithms is not the logarithm of the sum!). Hence, depending on the (completely arbitrary) choice of the vertical axis minimum, they will give very different and very misleading results (Figure 6-7c and 6-7d).

Don't even think of making a bar plot in the logarithmic scale.

You should pay attention to the dynamic range of a bar plot. If data shown are highly variable and cover a wide range of values, the

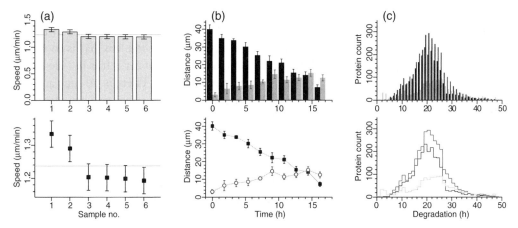

Figure 6-8. Inadvisable use of bar plots (top panels) and corresponding better alternatives (bottom panels). (a) The range of speeds is very compressed and unreadable in the bar plot. (b) Two overlapping bar plots are less clear than a scatter plot. Also, bar plots should not be used in examples (a) and (b) because the quantity shown is not additive. (c) There are three distributions of protein degradation time, coming from three cellular compartments, shown in one graph. Again, overlapping bars are less clear than line histograms shown in the bottom panel.

space available in the plot is effectively used. However, if your data vary only a little, you might end up with very similar bars, as shown in Figure 6-8a. This is because the bar plot forces the vertical axis to start at zero. When using a scatter plot as in the lower panel, the range of the vertical axis can be adjusted to present data clearly.

Bar plots are not ideal for presenting data with small variability.

Another problem with bar plots is that they become incomprehensible if you try to present several data sets with many points. This is demonstrated in Figure 6-8b and 6-8c. The plot in Figure 6-8b shows randomly generated data, but it mimics an actual plot I saw in a seminar talk. It looks very busy and breaks the main rule of clarity in making plots. The scatter plot version in the bottom panel is much easier to read.

Moreover, multiple-data bars make this plot ambiguous. When the horizontal axis is not categorical (as in this case), the bar width usually indicates the range over which data were collected for this bar. For example, in Figure 2-7a, radioactive decay events were counted over 1-s intervals, and each bin covers 1 s horizontally.

This doesn't work when you have more than one category, as demonstrated in Figure 6-8b. Look at the first point on the left (in both data sets). In the upper panel, the black bar is to the left of zero, and the grey bar is to the right of zero. It is not clear at all to what time these two bars correspond. There is no such ambiguity in the scatter plot in the lower panel, where you can clearly see that the first point corresponds to the time of zero, in both data sets.

> Multiple data bar plots are not suited for plots where the horizontal axis is not categorical.

Figure 6-8c shows a frequency distribution of protein degradation from three data sets. Bar plots, as shown in the top panel, do not work very well, as the resulting plot looks very cluttered. We could use colour to improve readability of the plot. However, it still suffers from the same problem as panel Figure 6-8b: it is not clear which values on the horizontal axis a given bar corresponds to. The bottom panel shows the same histograms but is drawn using lines with different colours (shades of grey, in this case). These are not perfect, and they require some effort to read, but line histograms are clearer than the overlapping bar plots. Here, the width of each histogram 'bar' (a horizontal segment) corresponds exactly to the range of x-values over which the proteins were counted.

> Multiple data bar plots can be cluttered and unreadable.

Unfortunately, this is not the end of the rant. Figure 6-9a shows a kind of plot I see all too frequently in the literature. Some people call it a dynamite plot[3]. It is supposed to compare a proportion of a quantity (e.g. cell number), in two treatments (called T1 and T2), with respect to the wild type (WT). These two values have errors, whereas the WT is not a measurement and is plotted only for reference. Such a graphical setup is common in biological publications.

There are a couple of things not quite right with this plot. First of all, error bars are unidirectional. The lower arm of each error

---

[3]Search the internet for 'dynamite plots' and see more criticism. It is not only me fulminating about them!

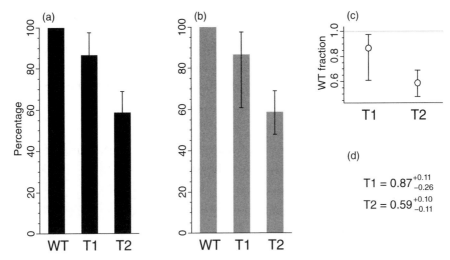

Figure 6-9. Bar plots and error bars. (a) 'Dynamite plot'; only upper error bars are visible. (b) Errors are not necessarily symmetric! (c) Quite often, it is better to show symbols with error bars than a bar plot. The fixed 100% WT reference is not a data point. (d) With only two numbers, a graph might not be necessary.

bar is not visible. The reader can only guess that errors are, perhaps, symmetric and extrapolate the lower error from the upper bit.

A simple change in colours solves the problem, as demonstrated in Figure 6-9b. In this particular example, error bars are asymmetric (and confidence intervals of a proportion usually are asymmetric!), so our initial impression of how T1 and T2 are related can be completely wrong. It seems very obvious that error bars should be fully visible, but for some reason (perhaps due to default settings in the graphical software used) this mistake is rather common.

Confidence intervals of a proportion are discussed in Section 5.8.

> Make sure that both the upper and lower arms of each error bar are clearly visible.

In my personal opinion, this plot can be further improved. Because all proportions shown are rather high, the plot has to be stretched vertically in order to show the results clearly. Using symbols instead of bars (Figure 6-9c) allows us to rescale the vertical axis and make the plot more compact. In addition, the 100% WT bar is not a measurement, so it is not necessary to show it as a data point. Instead, I have added a horizontal line at the fraction of 1 for easy comparison. The resulting plot is cleaner and easier to read and it saves some space, which might be important in a figure-heavy article.

But do we always need a plot? A picture might be worth a thousand words, but if you can present the result in just two numbers with their errors, then perhaps the plot is not necessary at all (Figure 6-9d).

> If you want to present only two numbers, a plot is not always the best solution.

Pie charts

Err … no. Pie charts are an awful way of presenting any data. They are uninformative and confusing. It is very hard to compare visually two sections of a pie chart with similar area. Data from any pie chart can be presented in a bar plot to a much better effect. Besides, you cannot add error bars in pie charts! Don't do pie charts. Just don't.

Overlapping error bars

Treatment–control comparisons are all in a day's work for a biologist. Often, this includes two sets of data (two samples). There exist statistical tests to compare various aspects of two (or more) samples and tell us if they are different. For example, to find out whether the sample mean is significantly different between two conditions, we might use a t-test. The null hypothesis is that both samples come from populations with the same mean. A two-sample t-test would tell us the probability that the observed difference between the means is due to random sampling. When the test p-value is small (e.g. < 0.01) we can reject the null hypothesis, while the p-value represents the risk of rejecting a true hypothesis. However, if the p-value is large, we can't say anything. This is where statistics turns nasty and shows us a finger. You cannot claim that the means are the same.

For hypothesis testing (not covered by this book), you can refer to standard textbooks, (e.g. Sokal and Rohlf 1995).

This is what statistical tests are for, and any sample comparison should be accompanied by the appropriate test p-value. Every now and then, you might find a published figure with data points and error bars (see Figure 6-10a) but no p-value immediately available. In such cases, you might be tempted to guess if the two values are significantly different or not, based on whether their error bars

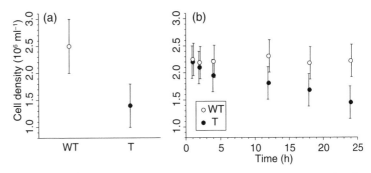

Figure 6-10. Overlapping error bars. (a) In some cases, overlap between error bars might be used as an indicator of whether the two conditions differ significantly or not. (b) Error bar overlap between two groups of data points does not have simple intuitive meaning. Note that error bars are horizontally staggered for clarity of presentation.

overlap or not. This is a risky business, and you should make sure that you understand what uncertainties the bars represent before you attempt the assessment. The following example shows the pitfalls of overlapping error bars. You can test your intuition on it.

Please read Exercise 6.2 at the end of this chapter, think about it for a while, make your choice and then read the solution in 'Solutions to Exercises' at the end of the book.

As you can see, standard deviations make the comparison very tricky. They don't take the sample size into account, so you need to do fiddly adjustments in your head. Standard errors are better, but even non-overlapping error bars do not guarantee that there is a significant difference between the samples. Probably the best option is to use 95% confidence intervals, as they naturally give the typically required significance when error bars do not overlap. But beware!

Non-overlapping 95% CI might indicate significant difference between sample means, but a proper statistical test should be performed to confirm this.

Looking at error bar overlap might help when comparing two conditions, each consisting of one sample or measurement. In other words, you might compare exactly two data points with errors in the plot. It is easy to imagine a situation where each condition consists of many samples, that is, several data points (with errors) for each condition. For example, these points might represent a time course, as shown in Figure 6-10b. In such cases, overlapping error bars do not give information about the difference between the two conditions. Error bars might overlap for each pair of points, but the combined effect might still create a statistically significant difference between the conditions. For example, if you combine the last three data points in each condition in Figure 6-10b, their means are significantly different with $p = 0.007$.

6.3   When can you get away without error bars?

Normally, I would recommend plotting error bars in every plot. There are a few cases, however, where it is acceptable to skip error bars, and I am going to briefly discuss them in this section.

On a categorical variable

This is a rather obvious case, but I need to mention it for completeness. When one of the axes shows a categorical variable, for example a sample description (as in Figure 6-9c), or number (as in Figure 6-8a, lower panel), then there is no error on this variable.

When presenting raw data

Errors in biological experiments are usually found from statistical properties of a sample, for example its standard error. In such cases, either sample elements don't have any obvious uncertainties or else these uncertainties are small in comparison with the scatter of all data points (see the 'When errors are small and negligible' case below). When presenting individual points of these 'raw' data, there is no need to plot error bars. The scatter of data serves as a rough visual measure of error. Additional plot components can help the reader to see data scatter (e.g. box plots in Figure 1-1) or trend (e.g. the regression line with its confidence intervals in Figure 8-6).

Large groups of data points

Figure 6-3b (both linear and logarithmic plots) shows an example of thousands of data points plotted in one plot. It would be very difficult to plot all error bars together with data points, as it would make the plot horribly busy and unreadable. In such cases, we typically plot only data points with no error bars. The purpose of the plot is to show how data are distributed, rather than individual measurements. Each data point is shown in the context of others and their variability. In such plots, outliers are (sometimes) obvious. It might be a good idea to skip error bars for the bulk of data, but add them to a few selected outliers or points of interest. Please note that the 'large group' case overlaps with the 'raw data' case. Quite often, we have a lot of raw data points with no obvious uncertainties.

When errors are small and negligible

See Section 8.1 for explanatory and response variables.

Often one of the variables presented in a scatter plot is an explanatory variable, controlled by the experimenter. In such cases the uncertainty, although measurable, is going to be small. Figure 6-10b shows cell density plotted against time. The horizontal axis represents moments when samples were taken for analysis. This quantity is not measured but is fixed by the experimenter, although the distinction is a bit fuzzy. You still need to measure time, using a clock or a timer. Also, the exact moment of taking a sample is often not precisely defined. However, if samples are collected every few hours and it takes only a minute or so to do it, the error in time becomes negligible.

By *negligible*, I mean so small that it can be ignored in data analysis, as the other error, plotted on the vertical axis, is much more important. This is not always true; imagine an experiment in which samples are collected every five minutes but it takes two or three minutes to actually pick and process each sample. In such cases, the error of time should be estimated by the experimenter and included in the plot.

Occasionally, errors in both axes can be so small that they are invisible in the plot. In such cases error bars can be skipped, but a comment stating that 'errors are smaller than symbols' is mandatory in the figure caption or legend.

Where errors are not known

In an ideal world, the basic rule is simple:

If you don't know the uncertainty of your result, go back to the lab and repeat the experiment until you get it.

An example with these two numbers was shown in Chapter 1. A single number, with no uncertainty and no context (e.g. 19,086), is meaningless. If you plot it in a figure, it won't carry much meaning. In particular, it is impossible to compare it to another number (e.g. 39,361), and state that one of them is significantly greater than the other.

In the real world of laboratory practice, there are results that come with no uncertainties. Some modern instruments magically produce piles of numbers with no associated errors. There might be another quantity expressing confidence in the result, for example a *p*-value or a score. This doesn't help if you want to plot some of these numbers.

When presented with such a dilemma, you should consider two options. The first one is quite obvious: do the experiment in replicates and use the variability of each measurement (e.g. in the form of the standard error) to assess its uncertainty. This is by far the best way. If this is not possible, I would advise against plotting them in a figure unless there is a clear context of other numbers illustrating the distribution or variability of the data (see the 'When presenting raw data' and 'Large groups of data points' sub-sections above).

6.4 Quoting numbers and errors

There is one more thing I need to bring up before this chapter can be concluded. It is not exactly about plots and error bars, but it is closely related. Namely, how do you quote (or report) numbers with their errors in writing? It seems a very straightforward business: you write the best estimate of the quantity in question followed by a plus-minus sign and the corresponding error. Where appropriate, you add units and get, for example, 1.5 ± 0.3 µm. Simple as pie. Or is it? In fact, it is quite surprising to see how many people get it wrong in publications, even in reputable journals. Clearly, there is more to presenting numerical results than just writing, or typing, some digits.

Significant figures

Firstly, we have to go back to basics and revise *significant figures*.

> Significant figures (or digits) are those that carry meaningful information.

This implies that the remaining digits that may be present beyond the significant figures are random junk. In most cases, they are by-products of calculations carried out to higher precision than that of the original data. Also, a measuring device can report more digits than its actual precision. You should never, ever quote them in a publication.

See discussion of reading errors in Section 3.5.

Consider the following example. A microtubule has grown 4.1 μm in 2.6 minutes. What is the speed of growth of this microtubule? If you divide 4.1 by 2.6 in a pocket calculator[4], you will get 1.5769230775 μm/min. However, time and distance are recorded down to only two significant figures, and their actual accuracy can be even worse. Intuition tells us that only the first two digits of the obtained number have any real meaning. The remaining figures are a by-product of the calculation and are completely meaningless. We can only *estimate* that the speed of growth is about 1.6 μm/min, but even this is not quite certain until we get more measurements (replicates) and find the speed uncertainty.

Writing significant figures

There are simple conventions stating which digits of a quoted figure are significant.

> Generally, all digits, except leading and trailing zeroes, are regarded as significant.

For example, 1.893 and 365 have four and three significant figures, respectively. There are, however, exceptions from this rule.

> Trailing zeroes after the decimal dot are considered significant.

0.0034 has two significant figures, but writing the same number down as 0.003400 indicates that the last two zeroes are also significant (i.e. we are confident in the last four digits: 3, 4, 0 and 0).

---

[4]A note to younger readers: in the old days a pocket calculator was a dedicated electronic device performing a similar function to a 'calculator' app on your mobile phone.

[5]To be more precise, the result is a periodic number 1.5(769230).

The confidence in the last significant figure is not always so great, and should be expressed by a corresponding error. I will get to it in a moment.

Trailing zeroes in integer numbers can be ambiguous.

Leading zeroes, on the other hand, are never significant. 34, 0.34 and 0.0034 have two significant figures. The amount of information carried by these three numbers is the same: it is a number '34' and an order of magnitude.

If you say '4012 patients took part in the study' it is quite obvious that there were exactly four thousand and twelve patients. If you say '4000 people gathered on the market square' it probably means 'about 4000' people. This can be guessed from the context. However, the statement 'samples were taken from 4000 patients' can be ambiguous: it might be the exact number, but equally well it might indicate only a rough size of the sample. In such situations it might be better to quote an ambiguous number using exponential notation: 4×10^3 has one significant figure, while 4.000×10^3 has four. It is not a very elegant solution, but ambiguity should not be allowed in strict scientific writing. Alternatively, you can quote an approximate value as ~4000, which is usually interpreted as 'about 4000' (more about it later in this section). The examples discussed above are summarized in Table 6-2.

The last significant figure should be properly rounded in writing. The direction of rounding depends on the next digit, which

Table 6-2. Examples of quoted values with different significant figures (s.f.).

Value quoted	Number of significant figures
1.893	4
365	3
0.34	2
0.0034	2
0.003400	4
4000	1 or 4
4×10^3	1
4.000×10^3	4
4000.00	6

is the first non-significant figure. If it is between 0 and 4, the preceding digit stays the same; if it is between 5 and 9, it is rounded up. For example, if the result of calculations is 1.5234 and we know that only two figures are significant, it should be quoted as 1.5. A 1.5534 result would be quoted as 1.6.

> The last significant digit should be rounded.

Errors and significant figures

So, how do we find out how many figures of our number are significant? The best way is to get it from the error associated with the number. First, we need to find the error, whether it is a standard error, confidence interval or any other uncertainty, as discussed in previous chapters of this book. Then, we need to find how many digits of the error are significant. In most biological applications this will be one or two digits, so if you don't have any other way of finding out, it is safe to assume only one significant figure of the error.

For error in the error see equation (4-23).

When the number to quote is a sample mean, you can use error in the error to find the number of significant figures. Error in the error (introduced in Section 4.7) shows the level of uncertainty in the standard deviation, but the same formula can be used for the standard error or a confidence interval. Knowing how uncertain the error is, we can decide how many digits should be quoted. If the relative error in the error, $\Delta SE/SE$, is greater than 0.1, then the uncertainty of SE is at least 10% and thus we can only trust its first digit. For example, if the calculated standard error is $SE = 0.02377345$ and we find $\Delta SE/SE = 0.13$, then the true unknown value of the SE is somewhere between 0.018 and 0.029 (with ~68% confidence), so we definitely cannot trust anything beyond the first figure. Our best estimate of the standard error is then $SE = 0.02$.

Table 6-3 summarizes relative error in the error for various sample sizes and gives the suggested number of significant figures in the error that should be retained. Unless you have a really huge sample, you would typically keep only one or two digits of the error. Quoting too many figures in the error is a common mistake found in many publications. If you say, for example, that your result is 2.567 ± 0.165, you implicitly suggest that you used at least 10,000 replicates to find these numbers!

Table 6-3.   Error in the error, calculated using equation (4-23), for different sample size, n. The second column shows relative error in the error, $\Delta SE/SE$. The last column indicates the number of significant figures of the error that should be quoted.

Sample size	Relative error in the error	Number of s.f. to quote
10	0.24	1
100	0.07	2
1,000	0.02	2
10,000	0.007	3
100,000	0.002	3

Most biological experiments result in one or two significant figures in the error.

Once we have found the error in the error, we can truncate (and round) the corresponding number *at the same decimal place as its error*. Consider a sample of 10 measurements that give the mean of $M = 1.23457456$ and the standard error of $SE = 0.02377345$. For $n = 10$, we find that the error has only one significant figure ($SE = 0.02$) so we should truncate the mean at the second decimal place and quote $M = 1.23 \pm 0.02$. This is illustrated in Figure 6-11.

Table 6-4 demonstrates a few examples of correctly and incorrectly quoted numbers, with errors. The most common mistake, apart from quoting non-significant figures, is truncating the number at a different decimal place than the error. Example (e) shows how to quote either very small or very large numbers, using exponential notation.

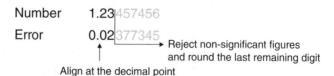

Figure 6-11.   Finding the number of significant figures from the error. Once the error in the error is known (see text), align the number and its error at the decimal point, and truncate the number at the same decimal place as you would truncate the error. Both numbers should be properly rounded at truncation.

Table 6-4. Examples of correctly and incorrectly quoted numbers with errors. In the 'incorrect' column, (a), (b) and (c) numbers are not truncated at the same decimal place as errors; (d) is formally correct, but a 5-digit precision of the error is highly unlikely in biological applications; and (e) is ambiguous due to missing brackets.

	Correct	Incorrect
a	1.23 ± 0.02	1.2 ± 0.02
b	1.23423 ± 0.00005	1.23423 ± 0.5
c	6 ± 3	6 ± 3.0
d	75000 ± 12000	75156 ± 12223
e	$(3.5 \pm 0.3) \times 10^{-5}$	$3.5 \pm 0.3 \times 10^{-5}$

Error with no error

Please read 'Where errors are not known' in Section 6.3 first. As I have explained before, you should make every effort to find the uncertainty of any quantity that makes its way to a publication. There are, however, many cases where errors are not very important in a particular context. Consider the following statements that might occur in published articles:

1. Centromeres are transported by microtubules at an average speed of 1.5 μm/min.
2. The new calibration method reduces error rates by ~5%.
3. Transcription increases during the first 30 min.
4. Cells were incubated at 22°C.

In all these examples, the numbers reported are only approximate. Although the exact value of the uncertainty is not known, a reader can at least guess the order of magnitude of the error. The rule of thumb is:

> All quoted figures are significant, and the uncertainty is in the last digit.

For example, we can presume that both digits of 1.5 μm/min are significant and the second of them is uncertain. We can *guess* that the error is probably not too big, for example ±0.2, or maybe ±0.3 μm/min. Otherwise, the authors would rather say '~1.5 μm/min' or 'around 1–2 μm/min'. We can also guess that the error is not too small. If it was ±0.01 µm/min, then, presumably, the authors would have quoted more figures, for example 1.52 µm/min.

The tilde (written as '~' and sometimes pronounced *twiddles*) usually means 'the same order of magnitude as'. The order of magnitude of a quantity is the number of powers of 10 in this quantity. For example, 5 is of the order of unity, so ~5 means somewhere between 1 and 10. 200 is in the hundreds, therefore, ~200 would mean somewhere between 100 and 1000. And so on. However, tilde is commonly used not to indicate the order of magnitude but rather to say 'roughly' or 'about'.

Statement (3) above usually makes more sense in a context. For example, it might refer to a figure showing how transcription rate varies in time. In such cases it is merely a comment on some other numerical data, and 30 minutes refers to a time interval that might make more sense with the figure.

The last example shows a number with two significant figures. We can suspect that the authors maintained the temperature within one degree of the stated value. If they said '22.4°C', we would expect much higher temperature accuracy. Again, the number of significant figures and the context of the statement should allow the reader to infer the likely extent of uncertainty.

Computer-generated numbers

Many biological experiments produce computer generated data in the form of a text file or Excel file. They typically consist of columns of numbers produced by the data-processing software. Usually these numbers are stored with excessive precision, way beyond the actual number of significant figures. The reason for this is to avoid the accumulation of rounding errors during calculations. You shall not quote all the digits generated by a computer in a publication or presentation. Truncate and round the quoted figure appropriately, using the rules laid out in this chapter (Table 6-6).

> Don't just copy and paste the computer output; always follow number-and error-quoting rules.

Another problem with computer-generated numbers is the so-called e-notation. A number 3.25×10^{-6} will be presented in computer output as 3.25e-6 or even 3.25e-006. It is quite disturbing to see how often this notation is used in publications: in tables, figures, even in the article text. This is horrible and should not be allowed! Instead of the dreadful 3.25e-006, you should use proper scientific notation, 3.25×10^{-6}.

> **Every time you use computer e-notation, a puppy dies.**

Finally, I have to mention a problem of fixed decimal places. This problem is specifically related to computer spreadsheets like Excel where it is rather hard to specify a given number of significant figures to display, but it is all too easy to specify a given number of decimal places. The left column in Table 6-5 is copied from a spreadsheet, where the formatting option was set to display one decimal place. As these numbers drop from thousands to less than one (presumably being logarithmically distributed), the number of displayed figures drops from 6 to 0. The last number is displayed as 0.0, but we don't really know whether it is a true zero, or something smaller than 0.05, rounded down to zero. This sort of presentation ought to be avoided. Assuming that there are two significant figures in each number, the right column in the table shows their correct representation. Some information is missing in the last two numbers, and it cannot be recovered without going back to the original calculations.

Table 6-5.   Fixed decimal places in a computer spreadsheet can produce the wrong number of significant figures. The left column is copied from a worksheet with one fixed decimal place. The right column shows the correct representation of these numbers (2 s.f.).

Wrong	Right
14524.2	1.5×10^4
2234.2	2200
122.2	120
12.6	13
2.2	2.2
0.1	0.1?
0.0	?

> Do not publish columns of values with a fixed number of decimal places; always round them to significant figures.

Summary

In this section, I have outlined the rules for quoting numbers and their errors in print. Although they sometimes might look like unnecessary formalities, they are crucial for clarity and correctness of presentation. These rules concern all sort of publications, reports, posters and presentations. They are summarized in Table 6-6. Please do follow them!

6.5   Exercises

Exercise 6.1

In a study of new antibiotics, bacterial cells were treated with three new compounds, called T1, T2 and T3. Viability of bacteria was studied in a resazurin assay, where the fluorescent intensity of each sample is proportional to the number of live cells. Each condition

Table 6-6.   Summary of rules for quoting numbers and errors.

When the error is known	When the error is not known

• Estimate the error in the error, using equation (4-23).
 • This will tell you how many significant figures of the error to quote (see also Table 6-3).
 • Typically, you quote 1–2 significant figures of the error.
 • Quote the number with the same precision as the error (see Figure 6-11).
	• You still need to guesstimate your error!
	 • Quote only figures that are significant, e.g. $p = 0.03$, not $p = 0.0327365$.
	 • Use common sense!
	 • Try estimating the order of magnitude of your uncertainty.
	 • Use '~' or 'about' qualifiers, where appropriate.
	 • Example: measure the distance between two spots in a microscope.
	    ○ Get 486.23 nm from computer software.
	    ○ Resolution of the microscope is 100 nm.
	    ○ Quote 500 nm.

(including the untreated wild type) was done in 12 replicates. The measured intensities (in arbitrary units) are as follows:

No.	WT ($\times 10^5$)	T1 ($\times 10^5$)	T2 ($\times 10^4$)	T3
1	9.08	2.86	1.24	33
2	16.5	3.73	1.33	63
3	11.7	1.95	1.52	51
4	3.11	1.76	0.989	39
5	27.4	12.5	2.74	71
6	1.24	1.29	0.514	36
7	8.86	2.2	1.34	67
8	11.2	3.55	1.89	84
9	0.972	0.735	0.669	22
10	4.48	2.09	1.07	42
11	2.21	1.10	0.519	676
12	3.22	1.12	0.383	954

Enter these numbers into a computer (I know, it's dull) and create a plot (using your favourite graphical software) to demonstrate the results. Think of all possible types of plots you could make. How would you represent data variability?

Exercise 6.2
Study Figure 6-12 carefully. The error bars in this figure are standard deviations. By looking at how much the error bars overlap, estimate whether the sample means in each pair are significantly different. Hint: make sure you take the sample size into account.

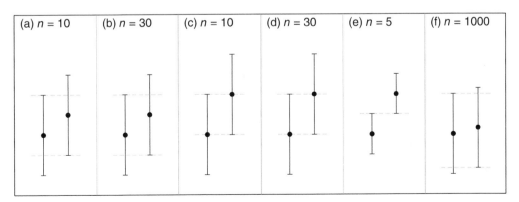

Figure 6-12. Overlapping error bars. Each panel contains two samples, consisting of n points each. Symbols and error bars show the mean and the standard deviation of each sample. See Exercise 6.2.

Chapter 7

Propagation of errors

If people do not believe that mathematics is simple, it is only because they do not realize how complicated life is.

—John von Neumann

In an experiment, two fluorescent dots are observed using a microscope. The instrument and its software provide the x, y and z coordinates of each dot. Due to the design of the microscope, resolution in the $x - y$ plane is better than in the z-direction. The measurement error for both the x and y coordinates is 120 nm, and the error of the z-coordinate is 200 nm. A postdoc makes a measurement of two fluorescent dots and records the following coordinates (in µm):

	x	y	z
Dot 1	3.68	3.12	5.44
Dot 2	3.90	3.86	4.02

Let us denote the coordinates of the first and second dots as (x_1, y_1, z_1) and (x_2, y_2, z_2), respectively. The corresponding errors are $\left(\Delta_{xy}, \Delta_{xy}, \Delta_z \right)$ for both dots, where $\Delta_{xy} = 0.12$ µm and $\Delta_z = 0.2$ µm.

Now we want to find the distance between the dots. The formula is $R = \sqrt{(x_2 - x_1)^2 + (y_2 - y_1)^2 + (z_2 - z_1)^2}$, and we find $R = 1.62$ µm. This is trivial. But, as you are very much aware, we are not satisfied by a result without any error. Hence, we have to find an uncertainty on the distance, ΔR. To do this, we need to learn how to *propagate* errors from x, y and z into R. Propagation of errors is the topic of this chapter.

7.1 What is propagation of errors?

Experiments produce numerical values, but these seldom serve as final results. Usually, experimental data undergo various

Understanding Statistical Error: A Primer for Biologists, First Edition. Marek Gierliński.
© 2016 John Wiley & Sons, Ltd. Published 2016 by John Wiley & Sons, Ltd.

transformations. For example, we might be interested in a ratio of two conditions, hence we need to divide two numbers. Or, in order to present the results in a plot, we might want to take a logarithm of data. Generally speaking, data are often *transformed* through a function of one or more variables. The logarithm is a function of one variable, $y = \log x$, and it transforms variable x into a new variable, y. The ratio is a function of two variables, $y = x_1/x_2$, and two variables, x_1 and x_2, are transformed into a new variable y. I'm going to use general notation to indicate transformation of one variable,

$$y = f(x),$$

or many variables,

$$y = f(x_1, x_2, \ldots x_n).$$

While it might be trivial to transform variables in this way, it is not obvious how *errors* propagate from x to y. More specifically, if a quantity x has error Δx, what is the error, Δy, of $y = f(x)$, where f is a known (and simple) function? For example, $(3.1 \pm 0.5) \times 10^7$ bacteria were counted in a sample. We want to plot this result with a logarithmic scale. The base-10 logarithm of the count is $\log(3.1 \times 10^7) \approx 7.49$, but how do we find its error? Obviously, it is *not* $\log(0.5 \times 10^7)$!

> When you calculate new quantities from known values, you also need to propagate their errors.

Any type of error can be propagated: standard deviation, standard error or a confidence interval. However, error propagation is most often applied to instrumental errors, when a new quantity has to be derived from measurements. You can propagate uncertainties derived from replicates, but see Section 7.4 for more discussion.

7.2 Single variable

Consider the following problem. We want to transform a value x into a new value y using a formula $y = f(x)$, where f is an analytical function of x, for example a logarithm. If x has uncertainty Δx, what is the uncertainty Δy of y? Or, in other words, having given $x \pm \Delta x$, how do we find $y \pm \Delta y$?

The answer is simple; you need to multiply the uncertainty of y by the *derivative* of the transformation function,

$$\Delta y \approx \left| \frac{df}{dx} \right| \Delta x. \tag{7-1}$$

The derivation of equation (7-1) is in Section 7.7. I show the derivation of this formula at the end of the chapter. Although it requires some calculus, I urge the reader to study it, because apart from the maths it explains the intuitive *meaning* of error propagation (there are pretty figures to illustrate it!). I always think it is good to know where the given equation came from, instead of just using it blindly.

Have a look at a few particular applications of equation (7-1).

Scaling

I will start with the simplest case of scaling. If a quantity x is multiplied by a constant a, how does its error change? The transformation function in this case is

$$f(x) = ax,$$

with derivative

$$\frac{df}{dx} = a.$$

Plugging this into equation (7-1) gives us

$$\Delta y = |a| \, \Delta x. \tag{7-2}$$

The result is very simple and intuitive. The error scales by the same factor as the number. For example, 10 ± 1 scaled by factor 5 gives 50 ± 5.

Logarithms

Logarithms are used in experimental biology to express quantities that vary by several orders of magnitude. Consider a logarithm to an arbitrary base:

$$f(x) = \log_b x,$$

where $x > 0$. Its derivative is

$$\frac{df}{dx} = \frac{1}{x \ln b}.$$

Using error propagation, equation (7-1) gives us

$$\Delta y = \frac{1}{\ln b} \frac{\Delta x}{x}. \tag{7-3}$$

In the case of a commonly used base-two logarithm, we find

$$\Delta y = \frac{1}{\ln 2} \frac{\Delta x}{x} \approx 1.44 \frac{\Delta x}{x}.$$

For example, a fold change of $x = 4.0 \pm 0.5$ will transform into $\log_2 x = 2.0 \pm 0.2$. In the case of a base-10 logarithm, the constant in this equation is $1/\ln 10 \approx 0.43$. Hence, the bacterial count example from the beginning of the chapter will convert the count of $n = (3.1 \pm 0.5) \times 10^7$ into $\log_{10} n = 7.49 \pm 0.07$.

This result has an interesting property – a *relative* error in the linear space ($\Delta x/x$) translates into *absolute* error (Δy) in the logarithmic space. If you have data where errors are typically 30%, regardless of the measured value, then errors of logarithms will be approximately constant (Figure 7-1).

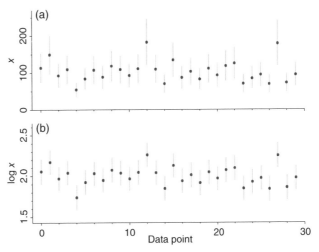

Figure 7-1. Errors of logarithms. (a) 30 measurements of x with uncertainties. Errors are proportional to measurements, $\Delta x/x \approx 0.3$. (b) Logarithms (base 10) of x and propagated errors. Errors proportional to x result in constant errors of $\log x$.

7.3 Multiple variables

Sometimes, multiple variables with their errors are transformed into a new quantity, and the transformation has a general form:

$$y = f(x_1, x_2, \ldots x_n),$$

where $x_1, x_2, \ldots, x_n$ are *independent* variables[1]. In most cases, when we need a sum or a ratio, there will be just two variables. But more complicated transformations are not unheard of. The general formula for error transformation is a simple extension of equation (7-1):

$$\Delta y^2 \approx \left(\frac{\partial f}{\partial x_1}\right)^2 \Delta x_1^2 + \left(\frac{\partial f}{\partial x_2}\right)^2 \Delta x_2^2 + \cdots + \left(\frac{\partial f}{\partial x_n}\right)^2 \Delta x_n^2,$$

$$(7\text{-}4)$$

The derivation of equation (7-4) is shown in Section 7.8. where $\partial f / \partial x_1$ denotes a *partial derivative* of f with respect to x_1 and can be understood as the *gradient* of function f in the direction of x_1. This formula looks a bit scary, but it is not difficult to derive. I show how to do it at the end of this chapter. Again, I encourage the reader to go through all the derivations and try to understand their *meaning*. In particular, I explain how *covariance* relates to this equation and why the variables $x_1, x_2, \ldots, x_n$ need to be independent.

Sum or difference

Let me show you how the general error propagation formula [equation (7-4)] can be applied in specific cases. First, let us consider a sum of two quantities, represented by the following function:

$$f(x_1, x_2) = x_1 + x_2.$$

---

[1]Very briefly, two random variables are independent when realization of one of them does not affect the probability distribution of the other. In practice, these are results of experiments that do not affect each other. For example, heights of two unrelated persons are independent, but heights of fathers and sons are not (see Exercise 5.3).

Partial derivatives of f with respect to x_1 and x_2 are

$$\frac{\partial f}{\partial x_1} = 1,$$

$$\frac{\partial f}{\partial x_2} = 1.$$

When used in equation (7-4) they give a familiar result,

$$\Delta y^2 = \Delta x_1^2 + \Delta x_2^2. \tag{7-5}$$

Errors add in quadrature. This makes perfect sense: errors are usually derived from standard deviations, and standard deviations of independent variables do add in quadrature. Errors of a difference $x_1 - x_2$ propagate in exactly the same way (i.e. they *add* in quadrature, not subtract!).

The result in equation (7-5) has a simple geometrical interpretation, shown in Figure 7-2a. The propagated error Δy can be understood as a hypotenuse of a right-angled triangle with legs Δx_1 and Δx_2. It can be seen that the propagated error is always greater than (or equal to) either of the original errors. Also, when Δx_1 is much larger than Δx_2, then $\Delta y \approx \Delta x_1$ and vice versa.

Ratio or product

Ratio is often used in biology to compare two conditions. The transformation function can be written as

$$f(x_1, x_2) = \frac{x_1}{x_2},$$

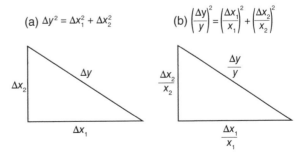

(a) $\Delta y^2 = \Delta x_1^2 + \Delta x_2^2$

(b) $\left(\frac{\Delta y}{y}\right)^2 = \left(\frac{\Delta x_1}{x_1}\right)^2 + \left(\frac{\Delta x_2}{x_2}\right)^2$

Figure 7-2. Geometrical interpretation of error propagation for (a) the sum or difference, and (b) the ratio or product.

and the derivatives are

$$\frac{\partial f}{\partial x_1} = \frac{1}{x_2},$$

$$\frac{\partial f}{\partial x_2} = -\frac{x_1}{x_2^2}.$$

Now we can put these into equation (7-4):

$$\Delta y^2 = \left(\frac{1}{x_2}\right)^2 \Delta x_1^2 + \left(-\frac{x_1}{x_2^2}\right)^2 \Delta x_2^2. \tag{7-6}$$

Assuming that $x_1 \neq 0$, we can multiply and divide the first term by x_1^2:

$$\Delta y^2 = \left(\frac{x_1}{x_2}\right)^2 \left(\frac{\Delta x_1}{x_1}\right)^2 + \left(\frac{x_1}{x_2}\right)^2 \left(\frac{\Delta x_2}{x_2}\right)^2$$

$$= \left(\frac{x_1}{x_2}\right)^2 \left[\left(\frac{\Delta x_1}{x_1}\right)^2 + \left(\frac{\Delta x_2}{x_2}\right)^2\right]$$

$$= y^2 \left[\left(\frac{\Delta x_1}{x_1}\right)^2 + \left(\frac{\Delta x_2}{x_2}\right)^2\right].$$

Hence,

$$\left(\frac{\Delta y}{y}\right)^2 = \left(\frac{\Delta x_1}{x_1}\right)^2 + \left(\frac{\Delta x_2}{x_2}\right)^2. \tag{7-7}$$

Relative errors add in quadrature. I will invite interested readers to prove that the product, $f(x_1, x_2) = x_1 x_2$, gives the same error propagation formula (see Exercise 7.1 at the end of the chapter).

Equation (7-7) is very similar to the propagation formula for the sum, except that relative errors replace absolute errors. The geometrical interpretation is the same, as shown in Figure 7-2b.

What happens if $x_1 = 0$? You can certainly calculate the ratio ($y = 0/x_2 = 0$), but how do you propagate errors? I was once supervising students doing a laboratory experiment in which they were taking a series of measurements, estimating their errors and calculating ratios. In this particular experiment, the numerator in the first measurement was always zero. They used equation (7-7) to propagate errors and invariably ran into a problem because x_1

was zero. They could calculate the ratio, but they were unable to propagate errors and usually returned a lab report with one error missing. When questioned about the missing value, they simply answered, 'I couldn't divide by zero, so I ignored it.' Surely, we can do better.

It is actually quite trivial. All we need to do is substitute $x_1 = 0$ in equation (7-6):

$$\Delta y^2 = \left(\frac{\Delta x_1}{x_2}\right)^2,$$

$$\Delta y = \frac{\Delta x_1}{|x_2|}. \tag{7-8}$$

The error of the ratio becomes insensitive to the uncertainty of the denominator, Δx_2. This doesn't mean that Δy remains small even if Δx_2 is huge. The error propagation formula works only when both errors, Δx_1 and Δx_2, are small, so the transformation function $f(x_1, x_2) = x_1/x_2$ is approximately linear when x_1 is disturbed by Δx_1 and/or x_2 is disturbed by Δx_2. Only in such cases equation (7-8) is correct for $x_1 = 0$. Obviously, we need $x_2 \neq 0$ all the time: indeed, you can't divide by zero.

I have collected error propagation formulae for a few commonly used transformations in the Appendix (Table A-3). You should exercise caution when using these equations, as they might not work in particular cases. If in doubt, always use the generic formula [equation (7-4)] and derive your particular case.

7.4   Correlated variables

As I stressed before, the general error propagation formula [equation (7-4)] is valid only when variables $x_1, x_2, \ldots, x_n$ are independent (uncorrelated). Otherwise, we need to use covariance terms (see derivation in Section 7.8). In the simplest case of two variables, the formula will look like this:

$$\Delta z^2 \approx \left(\frac{\partial f}{\partial x}\right)^2 \Delta x^2 + 2\frac{\partial f}{\partial x}\frac{\partial f}{\partial y}\text{Cov}(x, y) + \left(\frac{\partial f}{\partial y}\right)^2 \Delta y^2. \tag{7-9}$$

It is not always easy to find the covariance term, but I'm going to show an application of this formula to linear regression in Chapter 8.

7.5 To use error propagation or not?

Consider the following example. A test of a new drug in five repli-cated experiments gave the following values of half-maximal in-hibitory concentration (IC_{50}):

	R1	R2	R2	R4	R5
IC_{50} (nM*)	25	85	43	118	12
pIC_{50}	7.6	7.1	7.4	6.9	7.9

*Nanomolar is a unit of concentration, $1\ nM = 10^{-9}$ mol dm^{-3}.

The logarithmic version of IC_{50} is defined as $pIC_{50} = -\log(IC_{50}/M)$, where M is one molar. We want to find the mean and error of this quantity. We can do it two ways: (1) find the mean and standard error of IC_{50}, then convert it to a logarithm and propagate the error; and (2) find logarithms of each IC_{50} value and then find the mean and error. Let's try both methods. The mean and standard error of IC_{50} are $M = 56.6$ nM and $SE = 19.7$ nM, which can be written as

The rules for quoting numbers and errors are explained in Section 6.4.

$$IC_{50} = 60 \pm 20\ nM.$$

From this and the equation for error propagation, $\Delta y = 0.43\Delta x/x$, we can find

$$pIC_{50} = 7.2 \pm 0.1.$$

Alternatively, we can find logarithmic pIC_{50} values from individual measurements and their mean and standard error, $M_p = 7.38$ and $SE_p = 0.18$, which can be written as

$$pIC_{50} = 7.4 \pm 0.2.$$

These two results are slightly different. First of all, the logarithm of the mean does *not* equal the mean of the logarithm. As I stressed before, error propagation works well only when errors are small. In this example, the relative error of IC_{50} is 35% and the loga-rithm is not exactly linear at this scale. Hence, error propagation introduced a bias. The second method, where we calculated log-arithms of data first and then found the mean and the standard error, is better because the final result is unaffected by additional transformations.

Do not use error propagation if you can transform replicated data directly.

This advice usually refers to single-variable error propagation. With many variables, it is a bit more complicated.

Consider an experiment on mice where the effect of a drug on body mass is measured. We have a control group of 10 mice and a treatment group of 12 mice where the drug was administered. From a statistical test (e.g. a *t*-test), we know that the treated group is significantly heavier. Now we want to express this as a treatment-to-control ratio of body mass. We can do this by finding the mean and standard error of the body mass in each group, calculating the mass ratio and propagating errors using equation (7-7). We cannot calculate individual mass ratios for each mouse because the two groups consist of different mice! Not to mention that the two groups are not even equal-sized.

We can, however, imagine a slightly different experiment with just one group of mice. We weigh them, administer the drug for a certain period of time and weigh them again. Now we ask about the average relative mass gain after the drug was given. It makes more sense to calculate individual mass ratios and find their mean and standard deviation. But is it any better than finding the mean mass before and after exposure to the drug, calculating their ratio and propagating errors? There is no simple answer as it depends on what you want to measure. The first method would tell you the average body mass gain of an individual. The second method would *estimate* how the population mean had changed. They will give slightly different answers, and the choice of the answer is yours.

In practice, error propagation is used mostly when dealing with measurement (instrumental) errors, as in the example below.

7.6   Example: distance between two dots

Now, when we have gone through all the theoretical derivations, let us go back to the question posed at the very beginning of this chapter. I showed an example of an experiment, in which we found the 3D coordinates of two fluorescent dots and estimated their uncertainties. We want to find the distance between the dots and its error.

Once again, the distance between the two dots is

$$R = \sqrt{(x_2 - x_1)^2 + (y_2 - y_1)^2 + (z_2 - z_1)^2}. \tag{7-10}$$

Since we have six variables, x_1, y_1, z_1, x_2, y_2 and z_2, transformed into R, we are going to use the propagation formula for multiple variables (7-4). In order to do this, we need to find partial derivatives of R with respect to the coordinates. They are easy to find:

$$\frac{\partial R}{\partial x_1} = \frac{x_1 - x_2}{R},$$

$$\frac{\partial R}{\partial x_2} = \frac{x_2 - x_1}{R},$$

$$\frac{\partial R}{\partial y_1} = \frac{y_1 - y_2}{R},$$

$$\frac{\partial R}{\partial y_2} = \frac{y_2 - y_1}{R},$$

$$\frac{\partial R}{\partial z_1} = \frac{z_1 - z_2}{R},$$

$$\frac{\partial R}{\partial z_2} = \frac{z_2 - z_1}{R}.$$

Now we can use equation (7-4) to propagate errors:

$$\Delta R^2 = \left(\frac{x_1 - x_2}{R}\right)^2 \Delta_{xy}^2 + \left(\frac{x_2 - x_1}{R}\right)^2 \Delta_{xy}^2 + \left(\frac{y_1 - y_2}{R}\right)^2 \Delta_{xy}^2$$

$$+ \left(\frac{y_2 - y_1}{R}\right)^2 \Delta_{xy}^2 + \left(\frac{z_1 - z_2}{R}\right)^2 \Delta_z^2 + \left(\frac{z_2 - z_1}{R}\right)^2 \Delta_z^2$$

$$= \frac{2}{R^2}\left\{\left[(x_1 - x_2)^2 + (y_1 - y_2)^2\right]\Delta_{xy}^2 + (z_1 - z_2)^2 \Delta_z^2\right\}.$$

Finally,

$$\Delta R = \frac{\sqrt{2}}{R}\sqrt{\left[(x_1 - x_2)^2 + (y_1 - y_2)^2\right]\Delta_{xy}^2 + (z_1 - z_2)^2 \Delta_z^2}$$

$$\tag{7-11}$$

All we need is to use the numerical values from equations (7-10) and (7-11) and find $R = 1.62 \pm 0.26$ μm.

7.7   Derivation of the error propagation formula for one variable

Here I show one possible derivation of the error propagation formula for one variable, shown in equation (7-1). A quantity y is calculated from a known value x via a function $y = f(x)$. Having a known error Δx, we want to find the error Δy.

In general theory, x and y are random variables. For the purpose of this derivation, I'm going to apply a more intuitive approach and represent a random variable by a sample – a series of n measurements $x_1, x_2, \ldots, x_n$. The transformed measurements are $y_i = f(x_i)$ for all i. The mean and standard deviation of the sample x are M_x and SD_x, respectively. I will use standard deviations as a measure of errors, $\Delta x = SD_x$ and $\Delta y = SD_y$, although these can be easily replaced with either standard errors or standard deviation-derived confidence intervals. Our task is to find SD_y.

Have a look at a particular measurement x_i. Let $\delta x_i = x_i - M_x$ be the deviation from the mean (residual) of x_i. It is indicated on the horizontal axis in Figure 7-3. We can relate the propagated δy_i to the slope of function f, as illustrated in Figure 7-3a. When the curve representing f is almost flat, a given δx_i results in a small δy_i. In contrast, when f grows quickly, δx_j of the same size results in a much larger δy_j.

If deviations δx_i and δy_i are small, we can write

$$\frac{\delta y_i}{\delta x_i} \approx \frac{df}{dx}. \tag{7-12}$$

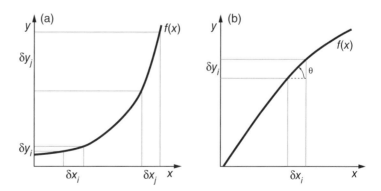

Figure 7-3.   Propagation of errors in the case of a single variable. The transformation is $y = f(x)$. (a) Deviation of the same size in x can create a very different deviation in y, depending on the slope of the curve, representing f. (b) The slope of f is its derivative, $\tan \theta = df/dx \approx \delta y_i/\delta x_i$.

The derivative is calculated at point x_i and it represents the slope of f ($\tan\theta$ in Figure 7-3b). In this equation, I replaced *differences* with *differentials*. This requires that δx_i and δy_i are small in comparison to the curvature of f (i.e. we request that within δx_i, the function f is approximately linear).

> Propagation of errors works well only when errors are small.

We can solve equation (7-12) for δy_i:

$$\delta y_i \approx \frac{df}{dx}\delta x_i. \tag{7-13}$$

Sample standard deviation is defined by equation (4-8).

Recall that δx_i and δy_i are small deviations from the mean (residuals). By definition, the variance (standard deviation squared) of y is

$$SD_y^2 = \frac{1}{n-1}\sum_{i=1}^{n}\left(y_i - M_y\right)^2 = \frac{1}{n-1}\sum_{i=1}^{n}\delta y_i^2. \tag{7-14}$$

We can replace δy_i with equation (7-13) and find

$$SD_y^2 \approx \frac{1}{n-1}\sum_{i=1}^{n}\left(\frac{df}{dx}\right)^2\delta x_i^2 = \left(\frac{df}{dx}\right)^2\frac{1}{n-1}\sum_{i=1}^{n}\delta x_i^2$$

$$= \left(\frac{df}{dx}\right)^2 SD_x^2.$$

In our intuitive approach, errors are represented by standard deviations, hence we can write

$$\Delta y^2 \approx \left(\frac{df}{dx}\right)^2\Delta x^2, \tag{7-15}$$

or, after taking the square root of both sides,

$$\Delta y \approx \left|\frac{df}{dx}\right|\Delta x. \tag{7-16}$$

The squared form is worth remembering, as it is consistent with the multiple-variable formula I derive in the next section.

7.8 Derivation of the error propagation formula for multiple variables

I will show how the error propagation formula is derived in the case of two variables. Calculations are less tedious than for n variables, but all the steps of the derivation are still present and the intuition behind it is the same. Let us call our two variables x and y, and the transformation formula is $z = f(x, y)$. This is a slight departure from the previous naming convention, but it will make equations easier to read.

Please read Section 7.7 first.

I will follow the same route as in deriving the single-variable formula in the previous section. x and y are random variables, represented by samples x_i and y_i, where $i = 1, \ldots, n$. From these numbers, we can find mean values M_x and M_y, and standard deviations, SD_x and SD_y. I'm going to use standard deviations as a measure of errors. Our problem is as follows: having SD_x and SD_y, what is SD_z?

Let $\delta x_i = x_i - M_x$ and $\delta y_i = y_i - M_y$ be deviations from the mean (residuals) for both samples. By analogy to equation (7-13), residuals of the resulting value, $\delta z_i = z_i - M_z$, can be expressed by the following formula:

$$\delta z_i = \frac{\partial f}{\partial x}\delta x_i + \frac{\partial f}{\partial y}\delta y_i, \quad \text{for } i = 1, \ldots, n \qquad (7\text{-}17)$$

The intuition here is similar to that in Figure 7-3. Each term in the sum consists of a gradient (derivative) multiplied by a residual. The gradient tells us how fast the function changes in the direction of x or y.

A one-dimensional gradient is illustrated in Figure 7-3b.

In one dimension, the gradient df/dx describes the slope of the curve $f(x)$. In two dimensions, $\partial f/\partial x$ is the slope of a two-dimensional surface $f(x, y)$ in the direction of x, and $\partial f/\partial y$ is the slope in the direction of y (see Figure 7-4). You can imagine this two-dimensional surface as a mountain. As you climb it, you can measure the slope along the north-south line and along the east-west line at each point. When multiplied by the corresponding residual, each gradient component gives us the contribution to the 'vertical' deviation in z. The sum of the two contributions gives the total deviation from the mean of z.

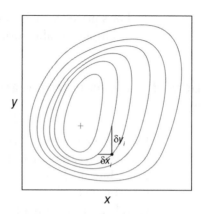

Figure 7-4. Illustration of a partial derivative in two dimensions. The grey contour plot represents a function $z = f(x, y)$. You can imagine this as a mountain presented in a map. From the point marked in the map, you can climb up by one contour, δz_i, by either moving in the x direction by δx_i, or moving in the y direction by δy_i. The partial derivatives with respect to x and y are then approximated by $\partial z/\partial x \approx \delta z_i/\delta x_i$ and $\partial z/\partial y \approx \delta z_i/\delta y_i$, respectively. Each derivative represents the slope of the mountain in the respective direction. In this example, the slope in x direction is much steeper than the slope in y direction.

Since we want to find the standard deviation of z, we need the squared residual first,

$$\delta z_i^2 = \left(\frac{\partial f}{\partial x} \delta x_i + \frac{\partial f}{\partial y} \delta y_i \right)^2$$

$$= \left(\frac{\partial f}{\partial x} \right)^2 \delta x_i^2 + 2 \frac{\partial f}{\partial x} \frac{\partial f}{\partial y} \delta x_i \delta y_i + \left(\frac{\partial f}{\partial y} \right)^2 \delta y_i^2.$$

We use δz_i^2 to find the variance of z,

$$SD_z^2 = \frac{1}{n-1} \sum_{i=1}^{n} \delta z_i^2 = \left(\frac{\partial f}{\partial x} \right)^2 \frac{1}{n-1} \sum_{i=1}^{n} \delta x_i^2$$

$$+ 2 \frac{\partial f}{\partial x} \frac{\partial f}{\partial y} \frac{1}{n-1} \sum_{i=1}^{n} \delta x_i \delta y_i + \left(\frac{\partial f}{\partial y} \right)^2 \frac{1}{n-1} \sum_{i=1}^{n} \delta y_i^2$$

$$= \left(\frac{\partial f}{\partial y} \right)^2 SD_x^2 + 2 \frac{\partial f}{\partial x} \frac{\partial f}{\partial y} \mathrm{Cov}(x, y) + \left(\frac{\partial f}{\partial y} \right)^2 SD_y^2.$$

Pearson's correlation coefficient is introduced in Section 4.4.   The quantity $\mathrm{Cov}(x, y)$ in the middle term is called *covariance* and is related to Pearson's correlation coefficient [equation (4-15)] as $\mathrm{Cov}(x, y) = r SD_x SD_y$. When variables x and y are independent (uncorrelated), it covariance approximately zero. This is because the mixed term in the middle, $\delta x_i \delta y_i$, is sometimes positive, sometimes negative and its sum, $\sum_i \delta x_i \delta y_i$, is approximately zero. In contrast, the squared terms, δx_i^2 and δy_i^2, are always positive, so they add up to create positive variances.

See Section 7.4 for correlated variables.

For the purpose of error propagation, we assume that variables x and y are independent, so we can get rid of the covariance term. Otherwise, the covariance term *must* be included in calculations. This is a very important assumption.

Variables for simple error propagation must be independent.

If we interpret standard deviations as errors, we can rewrite the last equation (with the covariance term removed) as

$$\Delta z^2 \approx \left(\frac{\partial f}{\partial x}\right)^2 \Delta x^2 + \left(\frac{\partial f}{\partial y}\right)^2 \Delta y^2.$$

When extended to many variables, this gives the general error propagation formula (7-4). In the case of one variable, it reduces to equation (7-1).

7.9 Exercises

Exercise 7.1
We have two numbers and their errors: $x_1 \pm \Delta x_1$ and $x_2 \pm \Delta x_2$. We calculate their product, $y = x_1 x_2$. What is the error of y?

Exercise 7.2
The radius of a spherical cell was measured with 10% accuracy. When you estimate the volume of the cell from this radius, what is the accuracy of this estimation?

Exercise 7.3
Your task is to prepare a 10.0 mM solution of NaCl. The required accuracy of the concentration is 0.1 mM. Considering that you can measure the mass of NaCl to the nearest milligram and volume of the solution to the nearest millilitre, how much NaCl do you need to use to obtain the required molar concentration accuracy?

Chapter 8

Errors in simple linear regression

It is proven that the celebration of birthdays is healthy. Statistics show that those people who celebrate the most birthdays become the oldest.

—S. den Hartog

Regression analysis is a way of finding a relationship between two or more variables. For example, we might be interested in how body mass depends on height, or how microbial cell number changes with increasing concentration of a drug. In the simplest case of linear regression, which I'm going to discuss here, we want to fit a straight line to a set of points, represented by their coordinates x_i and y_i. I will show not only how to find the parameters of this line, but also how to estimate their uncertainties and evaluate the error of the prediction. At the end of this chapter, I will briefly mention a more general case of curve fitting.

8.1 Linear relation between two variables

A linear relation between x and y can be written as

$$y(x) = ax + b, \tag{8-1}$$

where a is the *slope* and b is the *intercept*. Unknown parameters a and b are found by *fitting* function $y(x)$ to data (x_i, y_i), $i = 1, \dots, n$. x and y are sometimes called independent and dependent variables, but this might create confusion with the concept of statistically (in)dependent variables. Therefore, I'm going to follow a different convention and call x the *explanatory variable* and y the *response variable*.

These terms reflect intrinsic asymmetry in data: we are interested in how y *responds* to changes in x, not the other way round.

Understanding Statistical Error: A Primer for Biologists, First Edition. Marek Gierliński.
© 2016 John Wiley & Sons, Ltd. Published 2016 by John Wiley & Sons, Ltd.

Sometimes x is a variable we actually control in order to measure the response. For example, we might test how cells react to a drug by changing its concentration over a certain range. In many cases, however, we don't control the explanatory variable, but we want to compare two quantities characterizing objects of interest, for example body mass and height (see the example below). In such cases, the mass can be regarded as a response to the height (the taller you are, the heavier you should be), modulated by all other factors: diet, exercise and so on.

> In linear regression, data are asymmetric: the response variable results from the explanatory variable.

Mean response

We can understand linear relation in the usual terms of population and sample. There is a (usually abstract) population of points with a true regression line:

$$\bar{y} = \alpha x + \beta. \tag{8-2}$$

Here, α and β are the population slope and intercept, and $\bar{y}$ is the *mean response* to x. Formally, we should discuss it in terms of a random variable Y, normally distributed around mean $\bar{y}$, for a given x. In real populations, we can approximate the mean response by averaging y over a small range of x.

Are you perplexed by h^2 in linear regression? Wait a few paragraphs.

Figure 8-1 shows data from a 1993 Hong Kong Growth Survey[1]: height squared, h^2, and body mass, m, of 25,000 adolescents. The straight line in Figure 8-1 represents linear regression calculated for the entire population using the least-squares method described later in this chapter. I have approximated the mean response by finding the average mass in horizontal bins containing 100 points each. The approximated $\bar{y}$ agrees very well with the regression line.

> The regression line follows the mean response in the population.

---

[1]Data obtained from http://wiki.stat.ucla.edu/socr/index.php/SOCR_Data.

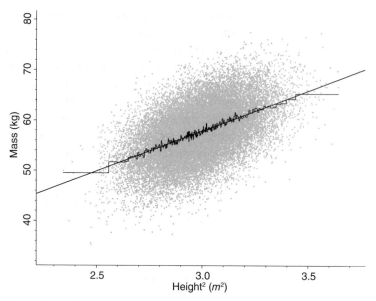

Figure 8-1. Population, regression and mean response. Grey points show data from the Hong Kong Growth Survey; the straight line is the regression line $m = \alpha h^2 + \beta$, where $\alpha = 15.9 \pm 0.3$ kg m^{-2} and $\beta = 10 \pm 1$ kg (95% CI). The histogram shows the mean mass in bins containing 100 points, and it approximates the mean response in the population data. Despite some variability, the mean response agrees with the population regression line very well.

True response and noise

Another remarkable thing about the data from Figure 8-1 is the substantial level of noise. Data points make a large splodge, and it looks almost like a miracle that the mean response line follows the straight regression line so closely. There is nothing miraculous about it. In fact, Figure 8-1 is telling us both that there is a true underlying linear relationship between m and h^2, and that the response is strongly affected by noise. The *true response* can be expressed as

$$y = \alpha x + \beta + r, \tag{8-3}$$

where r is the noise term. This equation reflects two things commonly found in biological data. Firstly, there is often a simple underlying pattern, for example a linear relationship. Secondly, it is obscured by noise, which can be very strong. This is how things are, and we cannot do much about it. The level of noise should not discourage you from analysing these data. Statistical tools like regression analysis enable us to extract the essence of the data and discover the underlying pattern.

> In biology, there is often an underlying simple relationship between observables, even if it is badly affected by noise.

Data linearization

By the way, I have plotted *squared* height on the horizontal axis in this figure in order to *linearize* the data. We expect m and h to be positively correlated: the taller you are, the heavier you are expected to be. However, body mass depends not only on height but also on, well, let's be frank: width. Physicians often use the body mass index (BMI), $B = m/h^2$, to characterize someone's bulkiness. Therefore, it is not unreasonable to expect a roughly linear relation between m and h^2. By transforming (h, m) into (h^2, m) I have converted data into their expected linear form, or linearized them.

Mind you, this is not just a random conversion: I used specific knowledge about the subjects of the study to do this. This procedure is recommended for all types of data. In particular, if we expect the response to have an exponential dependence on the explanatory variable, $y = ae^{bx}$, taking a logarithm of y will convert this into a linear relation between x and $\log y$: $\log y = \log a + b \log(e) \times x$. Likewise, converting both x and y into their logarithms linearizes a power-law dependence, $y = ax^b$, into $\log y = \log a + b \log x$.

> Linearize your data where possible.

Alas, we cannot linearize an exponential relation with an offset, $y = y_0 + ae^{bx}$. In such cases, we need to use a more general curve-fitting procedure; see Section 8.6 at the end of this chapter.

8.2 Straight line fit

The concept of statistical estimator was introduced in Section 4.2.

From the population we go down to a sample (x_i, y_i). Parameters a and b we want to find are statistical estimators of true unknown parameters α and β in the same sense as the sample mean, M, is an

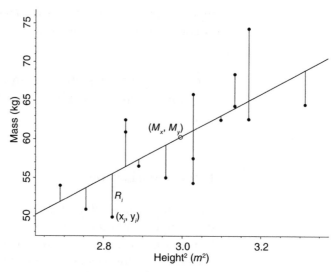

Figure 8-2. A sample of 16 points randomly drawn from the Hong Kong Growth Survey data presented in Figure 8-1. The model shown is the best-fitting straight line $m = ah^2 + b$. The residuals – deviations of data from the model – are shown as grey vertical lines. The open circle indicates the centroid of the data, (M_x, M_y).

estimator of the unknown population mean, μ. From the point of view of the sample, the *actual response* is

$$y_i = ax_i + b + R_i, \tag{8-4}$$

Residuals defined as the deviation from the mean are introduced in Section 4.4. where $R_i = y_i - y(x_i)$ is a *residual*. Please note that the term *residual* might have different meanings. It can be defined as either the deviation from the mean or, as here, deviation from the model. In either case, we calculate the *sum of squared residuals*, which is then used to find the sample's standard deviation [equation (4-8)], or to fit a straight line (below).

The prediction of the model for any given x, or the *predicted response*, is

$$y(x) = ax + b. \tag{8-5}$$

Figure 8-2 shows a sample selected from the Growth Survey data. The predicted model response (or, shortly, the model) is shown as a straight line. The residuals are shown as vertical grey lines: deviations of data from the predictions of the model.

So, how do we find our model parameters in practice? We want data points to lie as close to the fitted line as possible. This is usually done by minimizing the spread of residuals. Since the residuals can be either positive or negative, the best approach to straight line fitting is to minimize the sum of squared residuals,

$$Q = \sum_{i=1}^{n} R_i^2 = \sum_{i=1}^{n} (y_i - ax_i - b)^2. \tag{8-6}$$

This is called the *least-squares method*.

> The least-squares method minimizes the sum of squared residuals.

The condition for Q to be minimal is that partial derivatives with respect to both model parameters are zero: $\partial Q/\partial a = 0$ and $\partial Q/\partial b = 0$. After doing a bit of (rather straightforward and boring) calculus, we can solve these equations for a and b:

$$a = \frac{S_{xy}}{S_{xx}},$$
$$b = M_y - aM_x. \tag{8-7}$$

I use the following notation:

$$M_x = \frac{1}{n} \sum_{i=1}^{n} x_i,$$

$$M_y = \frac{1}{n} \sum_{i=1}^{n} y_i,$$

$$S_{xx} = \sum_{i=1}^{n} (x_i - M_x)^2, \tag{8-8}$$

$$S_{yy} = \sum_{i=1}^{n} (y_i - M_y)^2,$$

$$S_{xy} = \sum_{i=1}^{n} (x_i - M_x)(y_i - M_y).$$

The slope and intercept given by equations (8-7) are estimators, in the same sense as other statistical estimators discussed in Chapter 4. Our sample comes from a population, which is characterized by an unknown relation, equation (8-2). a and b are our best shots at estimating α and β. The good news is they are unbiased, so parameters averaged over many samples are equal to the true population parameters. Another interesting feature of the least-squares linear fit is that the best-fitting line always passes through the centroid of the data, (M_x, M_y), marked with the open circle in Figure 8-2.

8.3   Confidence intervals of linear fit parameters

Equations (8-7) show how to calculate the best-fitting slope and intercept of the least-squares straight line. Now we must find confidence intervals on these estimates. One way of doing this is by propagating errors from data to fit parameters.

Following asymmetry of explanatory and response variables, I assume that x is measured precisely and y is somehow uncertain.

See Section 8.6 for brief discussion of fitting data with known errors in. The problem is, uncertainties in y are not known. All we have here are 'naked' data points (x_i, y_i). However, errors in y can be derived from the spread of these data. We can use residuals (differences between data and the model) as a proxy for errors in the response variable. These can be directly propagated into errors of the fit parameters, a and b.

Residuals can be regarded as a manifestation of noise. They should be normally distributed around zero, and we can use the following estimator to calculate a standard deviation:

$$SD_R = \sqrt{\frac{1}{n-2}\sum_{i=1}^{n} R_i^2}.$$

(8-9)

See Section 4.8 for explanation of degrees of freedom. This is an unbiased estimator. The $n-2$ is here because we lose two degrees of freedom when calculating a and b. We can put $R_i = y_i - (ax_i + b)$ into this equation, and after a few simple transformations we find the form of SD_R useful in practical applications,

$$SD_R = \sqrt{\frac{S_{yy} - aS_{xy}}{n-2}}.$$

(8-10)

S_{yy} and S_{xy} are defined by equations (8-8).

SD_R represents the scatter of data points around the regression line, just like normal standard deviation represents the scatter of data around the sample mean. You can think of it as the standard

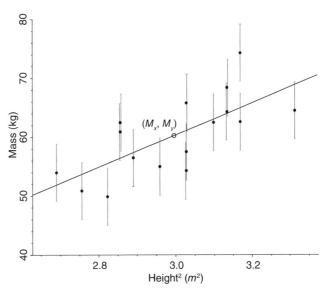

Figure 8-3. The sample from Figure 8-2 with the common error estimated from the standard deviation of the residuals, $\Delta y_i = SD_R$.

deviation of the response *after* eliminating the trend or correlation between x and y. SD_R tells us something about the level of
noise. It characterizes a *typical* error in the response. If we believe
that all measurements y_i are independent (they are not correlated),
and that the uncertainty of each y_i is similar, we can use SD_R as a
common uncertainty for each and every Δy_i:

$$\Delta y_i = SD_R, \tag{8-11}$$

for $i = 1, \ldots, n$. Our sample with common errors in y is shown in
Figure 8-3.

We can use standard deviation of the residuals, SD_R, to
estimate unknown errors in y.

Please note that SD_R cannot be used to assess the quality of the
fit. This is a very different situation to a case where errors Δy_i are
actually measured (or estimated) from the data. When we have
measured errors in the response variable, we can see how well our
data follow the model line, assess the fit quality (e.g. in terms of χ^2;
see Section 8.6) and decide if our model is correct. Here, we have
'naked' data points x_i and y_i with no error bars. We use the scatter

in the sample to find errors on fit parameters. We assume that the fit is 'good' and that the scatter of points around the model line is due to noise and not due to an incorrect model.

Fit parameters, a and b, can be derived from data points (x_i, y_i) using equations (8-7). Since there is an uncertainty in data, Δy_i, then there must be an uncertainty in the calculated fit parameters. We can use the general error propagation formula to find these uncertainties:

$$\Delta a^2 = \frac{SD_R^2}{S_{xx}},$$

$$\Delta b^2 = SD_R^2 \left[\frac{1}{n} + \frac{M_x^2}{S_{xx}} \right].$$

I show details of this derivation at the end of the chapter (Section 8.7).

The uncertainties Δa and Δb are propagated from common uncertainty in the response, which in turn is estimated by a standard deviation SD_R. Hence, Δa and Δb are also standard deviations. But standard deviations of what, exactly? Another *gedankenexperiment* can help us understanding this. Imagine we take a lot of samples from the population, and fit each of the samples with a straight line, finding its slope and intercept. From this, we can build the sampling distribution of the slope (shown in Figure 8-4) and the intercept (not shown). You can see a familiar trend in Figure 8-4: the smaller the sample size, the wider the sampling distribution and the slope is more uncertain.

I discuss sampling distribution in Section 5.1.

The width of the sampling distribution of the slope is estimated by our Δa (the same for the intercept). As you recall from Chapter 5, the width of the sampling distribution (its standard deviation, to be more precise) is the *standard error* of the corresponding estimator. Therefore, Δa and Δb are in fact standard errors of a and b, respectively,

$$SE_a = \frac{SD_R}{\sqrt{S_{xx}}},$$

$$SE_b = SD_R \sqrt{\frac{1}{n} + \frac{M_x^2}{S_{xx}}},$$

(8-12)

where S_{xx} is defined by equation (8-8).

Section 3.2 shows why we expect errors to be normally distributed.

In a typical situation, residuals (which can be regarded as random errors) are normally distributed. Under this assumption, the

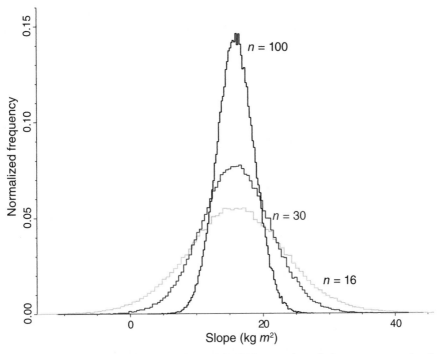

Figure 8-4. Sampling distribution of a slope. 100,000 samples of size *n* were randomly drawn from the Hong Kong Growth Survey, shown in Figure 8-1. For each sample, the slope and the intercept were calculated from a linear fit, using equations (8-7). The distribution of the slope for three sample sizes is plotted. The slope is normally distributed around the true value of $\alpha = 15.9$, with standard deviation estimated by equation (8-12).

estimators *a* and *b* are also normally distributed around true parameters characterizing the population (they are unbiased), as illustrated in Figure 8-4.

In Section 5.4, I showed a useful transformation [equation (5-1)] into a Student's *t*-distribution. It allowed us to find the confidence interval of the mean. Now we can use a similar trick,

$$t_a = \frac{a - \alpha}{SE_a}.$$

where *a* is the unknown true population parameter. The statistic t_a has a Student's *t*-distribution with $n - 2$ degrees of freedom. Following the method outlined in Section 5.4, we find the confidence interval for the slope,

$$a - t^* SE_a \leq \alpha \leq a + t^* SE_a, \tag{8-13}$$

where t^* is the critical value from t-distribution, corresponding to the assumed confidence level. A similar calculation can be done for the intercept:

$$b - t^* SE_b \leq \beta \leq b + t^* SE_b. \tag{8-14}$$

Here, β is the true unknown population parameter. The recipe for finding the best-fitting linear regression parameters with their errors is as follows:

1. Data points are (x_i, y_i), where $i = 1, \ldots, n$.
2. Use equations (8-8) to find M_x, M_y, S_{xx}, S_{yy} and S_{xy}.
3. Use equations (8-7) to find fit parameters a and b.
4. Use equations (8-12) to find standard errors SE_a and SE_b.
5. Find critical t^* from t-distribution with $n - 2$ degrees of freedom for the required confidence level.
6. Equations (8-13) and (8-14) give the confidence limits for the fit parameters.

Obviously, any self-respecting statistical software package will do this for you in a jiffy. But isn't it nice to know where all these numbers come from and what they really mean?

Example

Sixteen adolescent youths[2] have been measured and weighed. Their body mass (m) and height (h) are summarized in this table:

	1	2	3	4	5	6	7	8
h (m)	1.66	1.70	1.64	1.74	1.72	1.82	1.78	1.74
m (kg)	50.9	56.5	54.0	57.5	55.0	64.5	62.6	54.3

	9	10	11	12	13	14	15	16
h (m)	1.68	1.76	1.69	1.74	1.77	1.69	1.78	1.77
m (kg)	49.9	62.5	62.5	65.8	68.4	60.9	74.3	64.3

This sample is shown in Figure 8-2. We expect a roughly linear relationship between m and h^2:

$$m = ah^2 + b.$$

---

[2]The sample was randomly selected from the Hong Kong Growth Survey, see Figure 8-1.

Despite the square in h^2, this equation describes a linear model and can be fitted to (h_i^2, m_i) data points using the least-squares method. From equations (8-8), we find (4 s.f.)

$$M_x = 2.995 \text{ m}^2$$
$$M_y = 60.24 \text{ kg}$$
$$S_{xx} = 0.4439 \text{ m}^4$$
$$S_{yy} = 663.4 \text{ kg}^2$$
$$S_{xy} = 12.12 \text{ kg m}^2.$$

Then, equations (8-7) give us the slope and the intercept:

$$a = 27.30 \text{ kg m}^{-2}$$
$$b = -21.53 \text{ kg}.$$

The standard errors are calculated from equations (8-12):

$$SE_a = 7.314 \text{ kg m}^{-2}$$
$$SE_b = 21.94 \text{ kg}.$$

For sixteen data points and $n - 2 = 14$ degrees of freedom, the critical value from the t-distribution (significance level of 95%) is $t^* = 2.145$ (see Table A-1). Finally, our best-fitting values with their 95% CIs are

$$a = 27 \pm 16 \text{ kg m}^{-2}$$
$$b = -22 \pm 47 \text{ kg}$$

These errors are quite large. You can also see it from Figure 8-4, where the sampling distribution of the slope is very wide for $n = 16$. The actual standard deviation of this sampling distribution is about 7.6 kg m^{-2}, which is quite well approximated by our estimator $SE_a \approx 7.3$ kg m^{-2}. The main reason for large errors is that the sample doesn't span a large interval in h^2 and m, so it is not easy to establish the fit parameters accurately. In particular, when all data points are far from zero, the intercept is very poorly constrained and probably shouldn't be treated seriously. In this case, we might consider a different regression model. In the limit of the height of zero, we expect the body mass to be zero as well. Hence, we *expect* the intercept to be zero, and we should use regression through the origin (without intercept) instead (see Section 8-5).

It is worth noting that the slope, a, is *not* the BMI, despite having the same units. If we wanted to find the mean BMI of the group, we would have to calculate individual $B_i = m_i / h_i^2$ for each member and find their mean, standard error and the 95% CI. The result of this calculation is $B = 20.1 \pm 1.6$ kg m^{-2}, which is somewhat different from the slope.

8.4 Linear fit prediction errors

The best-fitting line of a simple linear regression gives us a prediction of y for every x, expressed by a simple formula $y(x) = ax + b$. In the previous section, we learned how to estimate errors of the fit parameters. Can we find the uncertainty of the prediction $y(x)$ itself? Since y is a function of a and b, $y = y(a, b)$, and errors (e.g. standard errors) of a and b are known, perhaps we could simply *Error propagation is* propagate errors from a and b to y, using the standard error prop-*discussed in Chapter 7.* agation formula [equation (7-4)],

$$SE_y^2 = \left(\frac{\partial y}{\partial a}\right)^2 SE_a^2 + \left(\frac{\partial y}{\partial b}\right)^2 SE_b^2.$$

Unfortunately, this is not going to work. The error propagation formula works only when the variables are independent. Parameters a and b are not independent! They are quite strongly correlated. This is quite easy to see from Figure 8-2. The best-fitting line always crosses the data centroid (the open circle). A change in the slope would cause 'pivoting' of the line about this centroid. An increase in slope (a steeper line) would decrease the intercept on the y axis and vice versa. This creates a strong correlation between the two variables. In order to illustrate this effect better, I have performed a simple numerical experiment. I have drawn 1000 random samples from the Growth Survey data, calculated the slope and the intercept for each sample and plotted these parameters in Figure 8-5. Clearly, there is a very strong (anti-) correlation between the fit parameters.

We need to take this correlation into account. The correct error propagation formula for correlated variables includes the covariance term and is given by equation (7-9). In our case, it takes the following form:

$$SE_y^2 = \left(\frac{\partial y}{\partial a}\right)^2 SE_a^2 + 2\frac{\partial y}{\partial a}\frac{\partial y}{\partial b}\mathrm{Cov}(a, b) + \left(\frac{\partial y}{\partial b}\right)^2 SE_b^2. \quad (8\text{-}15)$$

If residuals are normally distributed (and they usually are), we can calculate the estimator of covariance between a and b. It is given

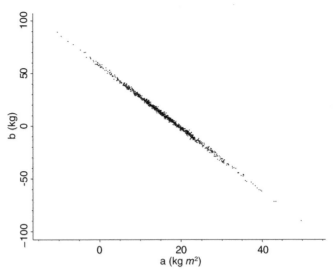

Figure 8-5. Illustration of correlation between the slope, a, and the intercept, b, of linear regression. 1000 samples of size $n = 16$ were randomly selected from the Hong Kong Growth Survey data. Each sample was fitted with a straight line, and the resulting fit parameters are plotted. The slope and the intercept are very strongly correlated, and this has to be taken into account when propagating errors.

by a simple formula (e.g. Press 2007),

$$\text{Cov}\,(a, b) = -SD_R^2 \frac{M_x}{S_{xx}}. \tag{8-16}$$

Taking into account that $\partial y/\partial a = x$ and $\partial y/\partial b = 1$ and substituting SE_a, SE_b from equation (8-12) and Cov (a, b) from equation (8-16) into equation (8-15), we get

$$SE_y^2 = x^2 \frac{SD_R^2}{S_{xx}} - 2x SD_R^2 \frac{M_x}{S_{xx}} + SD_R^2 \left(\frac{1}{n} + \frac{M_x^2}{S_{xx}} \right)$$

$$= SD_R^2 \left(\frac{1}{n} + \frac{x^2 - 2x M_x + M_x^2}{S_{xx}} \right) = SD_R^2 \left[\frac{1}{n} + \frac{(x - M_x)^2}{S_{xx}} \right].$$

Finally,

$$SE_y = SD_R \sqrt{\frac{1}{n} + \frac{(x - M_x)^2}{S_{xx}}}, \tag{8-17}$$

where SD_R and S_{xx} are given by equations (8-9) and (8-8), respectively. This formula gives the standard error of the prediction

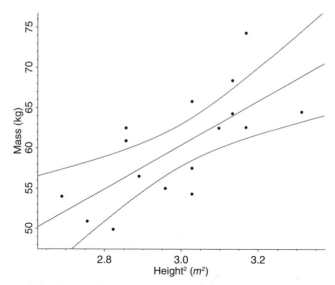

Figure 8-6. The same data and the best-fitting linear regression model, as in Figure 8-2. The grey curves represent the 95% confidence interval on the model prediction, found from equations (8-17) and (8-18).

$y = ax + b$ for any given x, where parameters a and b were calculated from a linear fit to the given data set (x_i, y_i). Because the prediction y is normally distributed, we can use the critical value from the t-distribution with $n-2$ degrees of freedom to find confidence intervals on y, as we did for the confidence interval of the mean and the confidence intervals of a and b,

$$y(x) - t^* SE_y \leq \bar{y}(x) \leq y(x) + t^* SE_y. \tag{8-18}$$

Let us look at the data from the 'Example' in Section 8.3, where we have already found the best-fitting line and the uncertainties of its parameters. Now, using equation (8-18), we can calculate the 95% CI on $y(x)$ for each x. The result, shown with grey lines in Figure 8-6, is rather typical; the error of the prediction is smallest in the centre, around the data centroid and gets larger towards the edges of the data. This is the effect of error propagation. If you disturb the fit parameters around their best-fitting values, the regression line will shift up and down a bit as a result of the varying intercept and will pivot slightly about the data centroid as a result of the varying slope. Effectively, it will trace a bowtie shape which can be seen in the figure.

8.5 Regression through the origin

The body mass and weight example in Section 8.3 shows that sometimes, due to the nature of the problem, we want to find a simpler relation between x and y, namely,

$$y = ax. \tag{8-19}$$

This is a model where the intercept is set to zero, known as *regression through the origin*, as the regression line is forced to go through the point $(0, 0)$. The mathematics here is a simpler version of derivations I presented previously, so I'm going to skip most of the details and only show where equations differ from the more general form.

The actual response is now $y_i = ax_i + R_i$, and by minimizing the sum of squared residuals we obtain

$$a = \frac{\hat{S}_{xy}}{\hat{S}_{xx}}. \tag{8-20}$$

This is equivalent to equation (8-7), except that the sums are redefined as

$$\hat{S}_{xx} = \sum_{i=1}^{n} x_i x_i,$$

$$\hat{S}_{xy} = \sum_{i=1}^{n} x_i y_i, \tag{8-21}$$

$$\hat{S}_{yy} = \sum_{i=1}^{n} y_i y_i.$$

The common uncertainty is, as before, the standard deviation of the residuals,

$$SD_R = \sqrt{\frac{\hat{S}_{yy} - a\hat{S}_{xy}}{n - 1}}. \tag{8-22}$$

The main difference between this formula and equation (8-10) is in the denominator. Here, we have $n - 1$, as we lose only one degree of freedom by calculating a (there is no b). Following the derivations shown at the end of this chapter, we can find the standard error of the slope,

$$SE_a = \frac{SD_R}{\sqrt{\hat{S}_{xx}}}. \tag{8-23}$$

Again, this is the same equation as in the case of the more general regression, except for the definitions of SD_R and $\hat{S}_{xx}$. The confidence intervals are found using equation (8-13).

The fit prediction errors are much simpler in the case of the regression through the origin. We have only one parameter, the slope a, and we can use the single-variable error propagation formula, given by equation (7-1):

$$SE_y = \left|\frac{dy}{da}\right| SE_a.$$

Because there is only one variable, we don't have to deal with the covariance term and we can apply this formula straightaway:

$$SE_y = |x| SE_a. \tag{8-24}$$

Recall that $dy/da = d(ax)/da = x$. The confidence interval can be found using equation (8-18) with the critical value from the t-distribution with $n - 1$ degrees of freedom.

Example

Let us redo the 'Example' from Section 8.3 using regression through the origin model. With the same data and equations (8-21), we get (4 s.f.):

$$\hat{S}_{xx} = 144.0 \text{ m}^4$$
$$\hat{S}_{yy} = 58730 \text{ kg}^2$$
$$\hat{S}_{xy} = 2899 \text{ kg m}^2.$$

Remember: the sums $\hat{S}_{xx}$, $\hat{S}_{yy}$ and $\hat{S}_{xy}$ are defined differently than in the previous case. Then, equation (8-20) gives us the slope:

$$a = 20.14 \text{ kg m}^{-2}.$$

The standard error is calculated from equation (8-23):

$$SE_a = 0.4056 \text{ kg m}^{-2}.$$

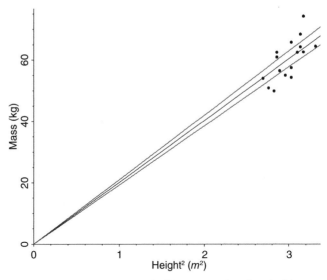

Figure 8-7. The same data as in Figure 8-6, but fitted with a straight line which is forced through the origin (see Section 8.5). The grey lines represent the 95% confidence interval on the model prediction, found from equations (8-24) and (8-18).

For 16 data points and $n - 1 = 15$ degrees of freedom, the critical value from the t-distribution (significance level of 95%) is $t^* = 2.131$ (see Table A-1). Finally, our best-fitting value of the slope with its 95% CI is

$$a = 20.1 \pm 0.9 \text{ kg m}^{-2}.$$

The uncertainty of the slope is an order of magnitude smaller than in the case of the free intercept. This is not surprising, considering the distribution of our data. As the points are clustered quite far from the origin, fixing the regression line at the origin removes most of the slope's uncertainty. This is illustrated in Figure 8-7, with the best-fitting line and its 95% CI. You can see that the regression line has much less room to pivot and the slope is constrained much better.

Before you use regression through the origin, make sure that this model makes sense with your data (i.e. that you really expect the response variable to be zero when the explanatory variable is zero). Otherwise, you might be applying the wrong model!

8.6 General curve fitting

Linear regression is a special case of curve fitting, where fit parameters and their uncertainties can be calculated analytically. In the

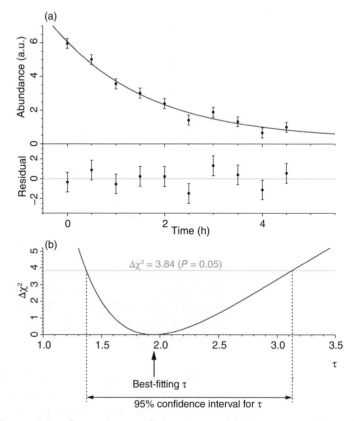

Figure 8-8. General curve-fitting errors. (a) An exponential model $f(t) = A_0 + Ae^{-t/\tau}$ was χ^2 fitted to 10 measurements, which are presented here with their 'one sigma' errors. (b) Finding confidence for the e-folding factor. τ was moved and fixed, and model refitted to find new χ^2. The 95% confidence limits on τ are marked by an increase in χ^2 by 3.84.

general case of arbitrary (non-linear) curve fitting, this is not possible. The subject of curve fitting is huge and complex and would require another book to explain things properly, so I'm going to refer the curious reader to the literature (e.g. Press 2007 or Motulsky and Christopoulos 2004).

Here I will briefly address the following problem: say, we have fitted a model (e.g. an exponential curve) to our data and obtained best-fitting parameters for this model (e.g. the e-folding factor). What are the uncertainties of these parameters?

Let us consider a very simple example, shown in Figure 8-8. We have n measurements y_i, with errors Δy_i taken at time points t_i. These are idealized, simulated data, but they can represent a real process of biological degradation. We want to quantify this process and find the time scale of decay and its uncertainty.

Mathematically, we can describe the exponential decay as a function of time:

$$f(t) = A_0 + Ae^{-t/\tau}. \tag{8-25}$$

Here A_0 is the asymptotic offset, A is the amplitude and τ is the time scale of the decay (sometimes called the *e*-folding factor). These three quantities are the fit parameters we want to find.

This exponential function can be fitted to the data, so it describes it in the best possible way. From the best-fitting model, we can find the best-fitting time scale, τ. The fitting is usually done so as to minimize the dispersion of data around the curve. The dispersion can be represented by the chi-square statistic,

$$\chi^2 = \sum_{i=1}^{n} \left[\frac{y_i - f(t_i)}{\Delta y_i} \right]^2. \tag{8-26}$$

It is not much different from the sum of squared residuals [equation (8-6)], but here each residual tells us how many error bars away the measurement is from the model. Residuals are shown in Figure 8-8a (bottom panel).

Note that Δy_i should represent standard deviations of data. Otherwise, χ^2 defined by equation (8-26) will not have the desired statistical properties. If your data points are sample means (which is often the case), then Δy_i should be standard errors (!) calculated *Standard error of the* from these samples. This is because the sample mean is normally *mean is explained in* distributed, and the standard deviation of this distribution is the *Section 4.5.* standard error of the mean. Simple?

Minimizing χ^2 is one of the least squares methods. There are clever ways of minimizing χ^2, and most of them require iterative numerical calculations. I used the Levenberg–Marquardt algorithm (Press 2007) to do the fitting. Most self-respecting statistical software packages will have some least-squares fitting methods built in. The best-fitting model parameters are $\tau = 2.0$ h, $A = 5.8$ and $A_0 = 0.25$. The minimized chi-square is $\chi^2_{min} = 6.92$.

Now we use the following procedure to find a 95% CI for τ. We are going to slowly move τ away from its best-fitting value and see what happens to χ^2. We do it by first decreasing τ in small steps. Each time we *fix* τ at a new value, allowing the two remaining parameters to vary freely, and refit the model to the data. With every new τ, the new minimized χ^2 is greater than χ^2_{min}, as we force the model away from the best fit. In other words, the fit becomes progressively worse as we move away from the optimum. We can continue pushing τ away from the best value until the increase in chi-square, $\Delta\chi^2 = \chi^2 - \chi^2_{min}$, reaches a certain critical limit. This marks our lower confidence limit, τ_L. Then, we can do the same on the other side of the best-fitting τ and find the upper limit, τ_U. The result is presented in Figure 8-8b.

It can be shown that $\Delta\chi^2$ is distributed as a chi-square distribution with one degree of freedom. Hence, from distribution tables we can find the critical value corresponding to the given confidence level. For $P = 0.05$ (95% CI), the critical value is $\Delta\chi^2 = 3.84$. The corresponding limits on τ are 1.4 and 3.1. Finally, our best-fitting time scale with 95% CIs is $\tau = 2.0^{+1.1}_{-0.6}$.

See Section 6.4 for rules on quoting numbers and errors.

8.7   Derivation of errors on fit parameters

Here I show the derivation of errors on linear fit parameters in the general case of regression with a free intercept. These fit parameters, a and b, are derived from data points (x_i, y_i) using equations (8-7), which I copy here for completeness:

$$a = \frac{S_{xy}}{S_{xx}},$$
$$b = M_y - aM_x.$$

We want to propagate uncertainty in the data, Δy_i, into uncertainties of the fit parameters, Δa and Δb. We use the general error propagation formula [equation (7.4)]:

$$\Delta a^2 = \sum_{i=1}^{n} \left(\frac{\partial a}{\partial y_i} \right)^2 \Delta y_i^2 = SD_R^2 \sum_{i=1}^{n} \left(\frac{\partial a}{\partial y_i} \right)^2,$$

$$\Delta b^2 = \sum_{i=1}^{n} \left(\frac{\partial b}{\partial y_i} \right)^2 \Delta y_i^2 = SD_R^2 \sum_{i=1}^{n} \left(\frac{\partial b}{\partial y_i} \right)^2. \tag{8-27}$$

The derivative of a with respect to y_i is

$$\frac{\partial a}{\partial y_i} = \frac{\partial}{\partial y_i} \frac{S_{xy}}{S_{xx}} = \frac{\partial}{\partial y_i} \frac{\sum_{k=1}^{n}(x_k - M_x)(y_k - M_y)}{S_{xx}}$$
$$= \frac{\partial}{\partial y_i} \frac{(x_i - M_x)(y_i - M_y)}{S_{xx}} = \frac{x_i - M_x}{S_{xx}}.$$

In the entire sum over k, only the ith term contains y_i, so derivatives with respect to y_i of the remaining terms are zero. We can find the derivative of b in a similar fashion,

$$\frac{\partial b}{\partial y_i} = \frac{\partial}{\partial y_i}\left(M_y - aM_x\right)$$

$$= \frac{1}{n}\frac{\partial}{\partial y_i}\sum_{k=1}^{n}y_k - M_x\frac{\partial a}{\partial y_i} = \frac{1}{n} - M_x\frac{x_i - M_x}{S_{xx}}.$$

Now we can substitute these two derivatives into error propagation equations (8-27),

$$\Delta a^2 = SD_R^2\sum_{i=1}^{n}\left(\frac{x_i - M_x}{S_{xx}}\right)^2$$

$$= SD_R^2\frac{\sum_{i=1}^{n}(x_i - M_x)^2}{S_{xx}^2} = SD_R^2\frac{S_{xx}}{S_{xx}^2} = \frac{SD_R^2}{S_{xx}}.$$

and

$$\Delta b^2 = SD_R^2\sum_{i=1}^{n}\left(\frac{1}{n} - M_x\frac{x_i - M_x}{S_{xx}}\right)^2$$

$$= SD_R^2\sum_{i=1}^{n}\left(\frac{1}{n^2} - 2\frac{M_x}{n}\frac{x_i - M_x}{S_{xx}} + M_x^2\frac{(x_i - M_x)^2}{S_{xx}^2}\right)$$

$$= SD_R^2\left[\frac{1}{n^2}\sum_{i=1}^{n}1 + \frac{2M_x}{nS_{xx}}\sum_{i=1}^{n}(x_i - M_x) + \frac{M_x^2}{S_{xx}^2}\sum_{i=1}^{n}(x_i - M_x)^2\right].$$

See equation (4-2) in Section 4.4. You might recall that the sum of deviations from the mean is always zero, hence the middle sum in the square brackets disappears. The first sum is simply n, and the last one was already defined as S_{xx}. Finally, we get

$$\Delta b^2 = SD_R^2\left[\frac{1}{n} + \frac{M_x^2}{S_{xx}}\right].$$

8.8 Exercises

Exercise 8.1
In an experiment, mRNA degradation was defined as a logarithm of the ratio of mRNA abundance at two different time points. Degradation was measured for 22 genes using Northern blots (x_i) and RNA sequencing (y_i). We want to compare these two methods. The mean and standard deviation of the two samples are as

follows (4 s.f.): $M_x = 0.3424$ and $SD_x = 0.4123$, and $M_y = 0.2805$ and $SD_y = 0.4187$. The correlation coefficient between x and y is $r = 0.8725$. Find the parameters a and b of a simple linear regression, $y = ax + b$, and their 95% CIs.

Exercise 8.2

Microbial growth was measured using dilution plating every 10 minutes. The estimated number of cells (in thousands) as a function of time (in minutes) is as follows:

Time	Count	Time	Count	Time	Count
10	2.2	50	10	90	50.1
20	2.6	60	17.0	100	60.4
30	5.7	70	19.5	110	79.3
40	7.8	80	29.5	120	92.1

Find a linear regression between time and count and its uncertainties. What is the culture doubling time?

Chapter 9

Worked example

Here we are. This is the end of the book. Well, almost. Before we finish, I'd like to show you one more example. This time, the example is going to be more elaborate and there will be more than one way of finding the final result. I will guide you, step by step, through all the calculations. Each derivation will involve various types of uncertainties and will refer you to different parts of the book. The data consist of only a handful of numbers and all calculations are very simple, so I encourage you to do them yourself.

9.1 The experiment

At the University of Southern North Yorkshire at Skipton, Prof. D. Nomal is conducting a study of a new cancer treatment drug called D42. He prepared a simple pilot experiment for his students to assess the effectiveness of the drug. He gave them a cancer cell line and asked them to compare cells treated overnight with the drug with an untreated control sample. Each of the students decided to employ a different strategy.

The first student, Sasha, divided the cells into six separate dishes, treated three of them with D42 and left the remaining three as a control. Thus, he created two biological conditions (treatment and control), each in three replicates. Sasha incubated the cells overnight, and the next day he took an aliquot from each dish and placed it in a counting chamber[1] under a microscope.

---

[1]A counting chamber, also known by the fancy name of *hemocytometer*, is a microscope slide with an indentation and a cover slip, to contain a small but precisely known amount of liquid. When a cell suspension is loaded into the chamber, one can count cells seen under the microscope.

Understanding Statistical Error: A Primer for Biologists, First Edition. Marek Gierliński.
© 2016 John Wiley & Sons, Ltd. Published 2016 by John Wiley & Sons, Ltd.

His intention was to count cells in the same volume of liquid and compare the numbers of living cells between the drug treatment and the control.

The second student, Lyosha, also wanted to use the counting chamber, but he decided to count both treated and untreated cells at the same time. Hence, he created two derivative cell lines, each expressing a different coloured fluorescent marker (green or red), and then treated the red cells with D42 while keeping the green cells as a control. He cultured them overnight in five replicates each. The next morning, he paired and mixed each replicate in equal proportions (i.e. replicate 1 from the control was mixed with replicate 1 from the treatment and so on). Then, he loaded a drop of each mixture into the counting chamber. This way he could count treated and untreated cells at the same time in the same volume, which, he hoped, would improve the accuracy of his results.

The third student, Masha, decided to go a different route and use a viability assay. She modified her cells to express a bioluminescent enzyme that degrades quickly in dead cells and hence can be used to quantify the amount of living cells. She prepared just two biological samples: the drug treatment and the control. She cultured them overnight, and the next morning she divided each sample into a row of a 96-well microplate[2]. Since a row of the plate consists of 12 wells, she obtained 12 technical replicates for both the treatment and the control. Then, she measured the intensity of light produced by each well, which was expected to be proportional to the number of living cells. These intensities were measured and recorded by an automated device which, after some data processing, produced a series of numbers (with arbitrary normalization). These numbers represent intensities of light emitted by the cells (not actual cell counts).

9.2 Results

The raw data obtained from these experiments are provided in Table 9-1 and visualized in Figure 9-1. All three experiments aim at finding the fraction of cells surviving the overnight treatment with D42. However, each of them does it in a different way.

---

[2]A microplate is a rectangular plate with small test tubes ('wells'). In this case, the plate has 12 rows and 8 columns of wells.

Table 9-1. Experimental results from the three students. Sasha and Lyosha counted cells, and Masha measured the intensity of a bioluminescent marker.

	(a) Sasha			(b) Lyosha				
No.	1	2	3	1	2	3	4	5
Control	111	104	123	93	84	117	101	112
D42	16	18	14	6	8	12	10	18

	(c) Masha					
No.	1	2	3	4	5	6
Control	411684	473908	339111	398736	465518	441436
D42	46479	44307	37646	52363	48027	44133
No.	7	8	9	10	11	12
Control	436648	416488	483275	374190	354678	406482
D42	41420	34763	89598	41936	38746	44930

Sasha

Let's start with Sasha's results. There are two conditions and three replicates. The replicates are independent of each other and are not matched or paired between the conditions. The simplest thing to do is to pool the replicates in each condition and add the cell

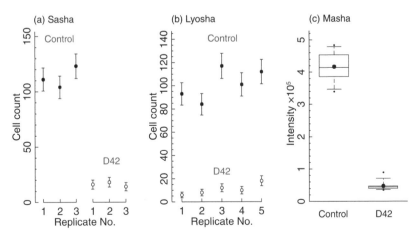

Figure 9-1. Illustration of raw data from Table 9-1 collected by the three students. The error bars in panels (a) and (b) are counting errors, estimated by the square root of the count. The box plots are as defined in Section 6.2; the large black spots represent the mean of each sample, and the small black symbols show the data in the top and bottom 5%.

counts. This gives us total counts of $n_c = 338$ and $n_t = 48$ (I'm going to use indices c for control and t for treatment). The fraction of surviving cells is defined as

$$f = \frac{n_t}{n_c},\tag{9-1}$$

and we obtain $f = 0.142$ from our counts. In this chapter, I will quote all the intermediate results with at least one non-significant figure at the end. This prevents the accumulation of rounding errors during calculations. In fact, I have carried out all the calculations in a computer, using the full available precision, just in case. The final results are always reported using only significant figures, following the rules outlined in Section 6.4.

Proportion and its uncertainty are discussed in Sections 4.4 and 5.8, respectively.   Let us find the uncertainty of f. We might be tempted to use the confidence interval of a proportion, but n_t is *not* a proportion of n_c. If you recall its definition, the proportion is a ratio of the number of 'successes' to the total number of 'trials'. The crucial point is that these successes are included in the trials: a *proportion* of the trials ended up with a success. Our surviving cells, n_t, are not included in the control, n_c. They are completely independent, and, if the drug is not particularly effective, it might even happen that $n_t > n_c$, which is impossible in a real proportion. Hence, the fraction f is not a proportion, in the sense defined by equation (4-16).

We can, however, find a genuine proportion here. We can 'pool' the treatment and the control together and find the proportion of the treatment survivals within the total count:

$$\hat{p} = \frac{n_t}{n_t + n_c}.\tag{9-2}$$

The binomial distribution is introduced in Section 2.7.   Using the counts from Sasha's experiment, we find $\hat{p} = 0.124$. This is a real proportion estimator and it should follow a scaled binomial distribution, as discussed in Section 4.4, so we can find its confidence intervals. Using equations (5-22) and (5-23), we find $p' = 0.128$ and $W = 0.033$. Hence, the 95% CI is between $p' - W = 0.095$ and $p' + W = 0.161$. This gives the lower and upper errors on $\hat{p}$ of $\Delta\hat{p}_L = 0.029$ and $\Delta\hat{p}_U = 0.037$, respectively. Following the rules of quoting the number and its error (Section 6.4), we can finally write $\hat{p} = 0.12^{+0.04}_{-0.03}$ (95% CI).

This is not what we are after, though, as we want f rather than $\hat{p}$, and these two quantities are different. Fortunately, it is very easy to derive a formula relating them, by comparing equations (9-1) and (9-2):

$$f = \frac{\hat{p}}{1 - \hat{p}}. \tag{9-3}$$

Error propagation is explained in Chapter 7. Now, since we have the uncertainty of $\hat{p}$, we can propagate it into f. The formula for a single-variable transformation (7-1) is

$$\Delta f = \left| \frac{df}{d\hat{p}} \right| \Delta\hat{p} = \frac{\Delta\hat{p}}{(1-\hat{p})^2}. \tag{9-4}$$

I will let you prove that $\frac{df}{d\hat{p}} = \frac{1}{(1-\hat{p})^2}$. Because errors of $\hat{p}$ are asymmetric, we need to find the error of f separately for the lower and upper errors of $\hat{p}$. Using equation (9-4), we find $\Delta f_L = 0.038$ and $\Delta f_U = 0.048$. Again, following the rules of quoting numbers and errors, we write $f = 0.14^{+0.05}_{-0.04}$ (95% CI).

This is *a* result. However, its uncertainty is based on the pooled data, and it assumes that there is an unknown true proportion of treatment survivals and that by random sampling the observed proportion $\hat{p}$ is binomially distributed. This doesn't take into account some possible additional uncertainties, such as biological variability and the non-uniformity of cell distribution in the test tube. As a result, the distribution of $\hat{p}$ *might* be wider than binomial and our uncertainty might be underestimated.

This is what we have our replicates for. The variability between replicates should include all additional sources of uncertainty and can be measured directly. Instead of pooling, it is better to average counts across replicates and find their uncertainties.

Confidence intervals of the mean are explained in Section 5.4. Let us use the sample mean and its confidence interval. The mean counts are $M_c = 113$ and $M_t = 16.0$, and their standard errors, from equation (4-20), are $SE_c = 5.55$ and $SE_t = 1.15$, respectively. The critical value from the t-distribution for the one-tail probability of 0.025 and 2 degrees of freedom is 4.303 (Table A-1). From this, using equation (5-5), we find the 95% CIs of $\Delta M_c = 24$ and $\Delta M_t = 5.0$. Then, we find the ratio of the means,

$$f_m = \frac{M_t}{M_c}, \tag{9-5}$$

which is $f_m = 0.142$ in our case. Since we know the uncertainties (confidence intervals) for the numerator and denominator, we can use the error propagation formula for a ratio, from equation (7-7):

$$\Delta f_m = f_m \sqrt{\left(\frac{\Delta M_t}{M_t}\right)^2 + \left(\frac{\Delta M_c}{M_c}\right)^2}.$$ (9-6)

Using the values derived above in this equation, we find $\Delta f_m = 0.053$. Hence, after rounding the result, we get $f_m = 0.14 \pm 0.05$. This result is similar to the one calculated from the pooled data, $f = 0.14^{+0.05}_{-0.04}$. This suggests that the additional errors I speculated about do not influence the experimental result very much. On the other hand, we have only three replicates, so our result (both the number *and* its error) is quite uncertain. I would suggest using more replicates to get a more robust outcome.

Lyosha

In many respects, Lyosha's experiment is similar to Sasha's and we can use the same approach to analyse his data. First, we pool the counts, calculate the proportion $\hat{p}$ and its confidence interval and then convert it into the fraction of living cells, f, and propagate its errors:

$$n_c = 507$$
$$n_t = 54$$
$$\hat{p} = 0.096$$
$$p' = 0.099$$
$$W = 0.025$$
$$\Delta\hat{p}_L = 0.022$$
$$\Delta\hat{p}_U = 0.027$$
$$\hat{p} = 0.10^{+0.03}_{-0.02}$$
$$f = 0.107$$
$$\Delta f_L = 0.027$$
$$\Delta f_U = 0.034$$
$$f = 0.11 \pm 0.03$$

Next, we look at the means across replicates with their uncertainties and the ratio, $f_m = M_t/M_c$. Here are all the intermediate steps and the final result:

$$M_c = 101$$
$$M_t = 10.8$$
$$SE_c = 6.04$$
$$SE_t = 2.06$$

$$t^* = 2.776$$
$$\Delta M_c = 17$$
$$\Delta M_t = 5.7$$
$$f_m = 0.107$$
$$\Delta f_m = 0.059$$
$$f_m = 0.11 \pm 0.06$$

Both approaches give the same result, but using replicates provides us with an uncertainty (95% CI) twice the size of that from the pooled data. This is because replicated experiments can capture all types of variability affecting the measured quantity at different stages of the experiment. I will discuss this in more detail in Section 9.3.

Lyosha's data consist of five *paired* replicates. Biological replicate 1 of the control was mixed in 1:1 ratio with biological replicate 1 of the D42 treatment, and then both types of cells were counted together in the same counting chamber. You might point out that there is little difference between this and counting cells in each sample separately (as Sasha did). You'd probably be right, as the counting process in this experiment is very simple. However, when the measurement itself is more complicated, the mark-and-mix approach can improve the results quite a bit. For example, in a type of mass spectrometry called SILAC[3] one cell sample (e.g. a control) is grown in a normal medium, while the other condition (e.g. a treatment) is grown in a medium containing heavy (but stable) isotopes in some of the amino acids (e.g. using heavy carbon ^{13}C instead of normal ^{12}C). When cells absorb the heavy carbon, their proteins become *marked* (labelled) in a way that can be distinguished by the mass spectrometer. Both samples are then mixed and processed together. Because mass spectrometry is a complicated procedure and introduces a lot of uncertainties, measuring the two samples in parallel reduces some of these errors significantly.

Lyosha's cell counting is, obviously, not very complicated, but it still might serve as an example of measuring marked cells together. As an alternative to the two ways presented above, we can calculate the surviving cell fraction separately for each matched replicate pair and then do the statistics. Individual fractions calculated from *Once again, confidence* the raw data (simply by dividing row 3 by row 2 in Table 9-1b) *intervals of the mean are* are as follows: 0.065, 0.095, 0.10, 0.099 and 0.16. The mean and *explained in Section 5.4.* standard error from these numbers are $f_p = 0.104$ and $SE_f = 0.016$,

---

[3]Stable isotope labelling by amino acids in cell culture.

and the critical t-value for 4 degrees of freedom is $t^* = 2.776$. This gives us the fraction from paired data, $f_p = 0.10 \pm 0.04$ (95% CI).

Because Lyosha's experiment consists of five biological replicates, it gives us a chance to study its reproducibility in more detail. If cells counted by Lyosha are randomly and independently distributed, they should follow a Poisson distribution. Hence, the count variance should be the same as the count mean. We can estimate the mean and the variance for each condition from the five replicates. From the raw data, we get:

Poisson distribution is discussed in Section 2.8.

$$M_c = 101$$
$$VAR_c = 182$$
$$M_d = 10.8$$
$$VAR_d = 21.2$$

In both conditions the variance is roughly twice that of the mean. This indicates that *perhaps* there is an additional component to the observed variability, beyond what we expect from random and independent counts. I say 'perhaps', because we deal with small numbers here and such excess in the variance can happen purely by chance. There are statistical tests to assess this. For example, Pearson's chi-square test[4] tells us that the probability of obtaining such a high variance purely by chance is about 0.1, for both control and D42 treatment counts. This is hardly a significant result, so we cannot conclude that the observed variance is really higher than the mean. We would need (many) more replicates to confirm this.

Correlation is explained in Section 4.4; its confidence interval and significance are discussed in Section 5.7.

There is one more thing we can do with these data. Since the replicates are (partially) paired, we might expect a certain level of correlation between the drug treatment and the control. The correlation coefficient, defined by equation (4-15), is $r = 0.742$. To find the confidence interval of this number, we use equations (5-17), (5-18) and (5-19). We get

$$Z' = 0.954$$
$$\sigma' = 0.707$$
$$Z'_{lo} = -0.432$$
$$Z'_{up} = 2.340$$
$$r_{lo} = -0.407$$
$$r_{up} = 0.982$$

---

[4] I don't want to go into details here, you can find more information about this test in online resources (e.g. in Wikipedia).

which gives a whopping uncertainty of the correlation coefficient, $r = 0.7^{+0.3}_{-1.1}$. Just in case, we can calculate the significance of the correlation. The t-statistic, defined by equation (5-20), is $t = 1.914$. With three degrees of freedom, this gives us the probability of obtaining this level of correlation by chance of $p = 0.08$. Just as in the case of variance, the result is not significant, and we conclude that the correlation coefficient cannot be determined due to insufficient data.

Masha

Masha's experiment doesn't give us cell counts but bioluminescent light intensities measured by a rather complicated device. Hence, we cannot assume anything about either the level of variability, or even the distribution of intensities. As you remember, some quantities in biological experiments can be log-normally distributed. In such cases, it might be better to log-transform all data before calculating statistics like the mean or standard error. However, a quick glance at Figure 9-1c shows that this is probably not the case here. The boxes in the plot are symmetric (the treatment box is less symmetric due to one outlier; I'll come back to this in a moment), whereas the log-normal distribution usually results in skewed boxes, as illustrated in Figure 6.3d. Hence, there is probably no reason for log-transforming these data.

Log-normal distribution is discussed in Section 2.6.

We can use the 12 replicates in each condition to find their means and confidence intervals, calculate a ratio and propagate errors. Before we do this, it might be a good idea to renormalize the data, just for convenience, so we don't have to handle large numbers. Let us divide both control and treatment intensities by the mean intensity from the control, $N_c = 416{,}800$. Then, we can follow the same steps as we did for Sasha's and Lyosha's data. This time we get

$$M_c = 1.000$$
$$M_t = 0.113$$
$$SE_c = 0.032$$
$$SE_t = 0.010$$
$$t^* = 2.201$$
$$\Delta M_c = 0.070$$
$$\Delta M_t = 0.022$$
$$f_m = 0.113$$
$$\Delta f_m = 0.023$$
$$f_m = 0.11 \pm 0.02$$

The mean control, M_c, is obviously 1, because of the way we normalized these data. The surviving cell fraction, $f_m = 0.11 \pm 0.02$, is consistent with the results obtained from the other data sets.

If you look at Masha's data carefully, you might notice that replicate 9 in the drug treatment looks suspiciously high, about twice the typical intensity in other D42 replicates. This 'outlier' might be due to either a random fluctuation or an actual problem in sample preparation and processing. For example, Masha could have made a pipetting mistake while loading samples onto the microplate. Unless we have a solid, independent proof that something has gone wrong, we cannot remove a datum just because it looks suspicious. In our case there is only one such point and, on its own, it does not affect the mean and its error very much.

Median is described in Section 4.4, and its confidence interval is discussed in Section 5.6. Alternatively, we can calculate a statistic that is not sensitive to occasional outliers, for example a median. I'm going to repeat the same steps as above, but instead of the mean and its confidence interval, I'm going to use the median and its simplified version of the confidence interval, described by equations (5-14), (5-15) and (5-16):

$$\widetilde{M}_c = 0.993$$
$$\widetilde{M}_t = 0.106$$
$$L_c = 4, U_c = 8, \widetilde{SE}_c = 0.036$$
$$L_d = 4, U_d = 8, \widetilde{SE}_d = 0.004$$
$$d.o.f. = 3$$
$$t^* = 3.182$$
$$\Delta\widetilde{M}_c = 0.12$$
$$\Delta\widetilde{M}_t = 0.011$$
$$f_{med} = 0.107$$
$$\Delta f_{med} = 0.017$$
$$f_{med} = 0.11 \pm 0.02$$

Well, the median and its error turn out to be very similar to the mean and its error. This is because the data are distributed symmetrically, so the mean and the median are essentially the same.

9.3 Discussion

Three independent experiments have been carried out to estimate the effectiveness of the new cancer drug, D42. The aim was to find out the fraction of cancer cells surviving a given dose of the drug after 12 hours. I have suggested various ways of analysing data from each experiment. The summary of results is shown in Figure 9-2.

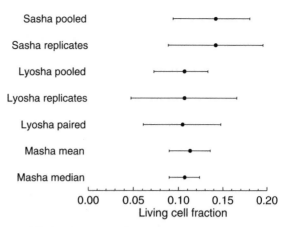

Figure 9-2. Results from various approaches to calculate the living cell fraction after D42 treatment. The experiments and the methods are described in the text. The error bars are 95% confidence intervals.

The good news is that all the results are consistent with each other within error bars (95% CI). The not-so-good news is that the error bars are quite dissimilar. For example, the confidence interval on Lyosha's replicated data is three times larger than the confidence interval on Masha's median. Even more worryingly, it is twice the size of the error bar of his own pooled data. The same experiment, depending on how we interpret its data, can give error bars of a very different size.

The problem with pooling data is that by adding counts, we lose crucial information about biological and technical variability. The error of a proportion reflects only the random distribution of counts between the two conditions (binomial distribution). In real life, there are many other sources of error: for example, due to the diversity of cells used in the experiment, variation in the initial cell numbers or density, conditions of growth in each test *See Section 3.2 for a* tube, uncertainty of the drug concentration, cell suspension non-*simple model of* uniformity or pipetting errors, to name just a few. All these errors *accumulated errors.* accumulate and contribute to the total uncertainty, which can be much more than the simple error of a proportion. This sort of uncertainty, although impossible to predict theoretically, can be estimated by using replicated data. And we *do* have replicates here, so it is better to use the information they contain, instead of pooling data.

If you have replicates, use them!

This is why errors from the replicated data are larger than errors from the pooled data. The replicates can 'see' all this additional variability, which the error of a proportion cannot.

Lyosha has paired his replicates in an attempt to reduce uncertainties from the counting chamber. Although this procedure resulted in a slightly smaller error bar, it is not clear whether the observed error reduction is due to the improved experiment or due to a random fluctuation. A correlation between the two conditions would suggest that the pairing is real and that uncertainties in the counting chamber (e.g. due to variation in the volume of the liquid) affect both types of cells in the same way. However, the correlation found was not statistically significant. Five replicates is really not enough to build convincing statistics!

> You always need more replicates.

Masha's experiment resulted in much smaller error bars than Sasha's and Lyosha's (replicated data); see Figure 9-2. It is not obvious why. We can only speculate that the small number of counts (in particular in the drug treatment) contributed to the larger error in Sasha's and Lyosha's experiments. Masha's measurement was based on a much larger sample (presumably millions) of cells, so *I discussed random and* her uncertainties are not count based. On the other hand, she *systematic errors in* didn't use independently cultured biological replicates, so her ex-*Section 3.1.* periment did not capture biological variability. Her random technical uncertainties came from inaccurate pipetting and noise in the bioluminescent marker activity. In addition, we should consider the possibility of a systematic error due to a nonlinear relation between the living cell number and the bioluminescent light intensity. We should also point out that Masha used 12 replicates in each condition, which certainly had reduced the random error.

9.4 The final paragraph

In this chapter, I wanted to demonstrate how various types of uncertainties are calculated in practice. Usually, biological data are complex and there is no one 'right' way of finding errors. Sometimes, a simple error of a proportion or correlation would suffice. But in most cases, independent replicates are needed to grasp the amount of intrinsic variability and estimate real errors. I hope this book will help you understand statistics a little bit more and use this knowledge to estimate uncertainties in real biological experiments. Remember: a measurement without error is meaningless.

Solutions to exercises

Exercise 2.1

	a	b	c	d	e
Distribution	Uniform	Log-normal	Poisson	Gaussian	Poisson
Mean	3.5	3.5	4	100	100
Standard deviation	0.87	0.90	2	10	10

Exercise 2.2

Firstly, we need to calculate the expected value, or the mean, number of functionally transfected cells. It equals the total number of cells multiplied by the transfection rate, $\mu = 3 \times 10^5 \times 10^{-5} = 3$. Since we are interested in the probability of having at least one transfected cell, $P(X > 0)$, the easiest way is to find the probability of having no transfected cells, $P(X = 0)$ first. Using the Poisson formula (2-11), we find

$$P(X = 0) = \frac{3^0 e^{-3}}{0!} = \frac{1 \times e^{-3}}{1} = e^{-3} \approx 0.05.$$

Hence, the probability of having at least one transfected cell is $P(X > 0) = 1 - P(X = 0) = 0.95$.

Exercise 2.3

You play the lottery buying one ticket a week. Hence, the probability of winning in the time interval of $\Delta t = 1$ week is $P_1 = 7.15 \times 10^{-8}$. We can solve equation (2-14),

$$P_1 = 1 - e^{-\mu \Delta t},$$

Understanding Statistical Error: A Primer for Biologists, First Edition. Marek Gierliński.
© 2016 John Wiley & Sons, Ltd. Published 2016 by John Wiley & Sons, Ltd.

for μ to find the mean winning rate:

$$\mu = -\frac{1}{\Delta t} \ln{(1 - P_1)}.$$

A year is, on average, 365.25 days long, which corresponds to 52.18 weeks. This gives $\Delta t = \frac{1}{52.18} = 0.0192$ years. Plugging this and P_1 into the above equation gives the mean winning rate of $\mu = 3.72 \times 10^{-6}$ per year. On average, you will have to wait $\frac{1}{\mu} = 269,000$ years to win the lottery. Good luck.

Exercise 2.4

Table A-1 shows right-tail probabilities for the t-distribution and Table 2-1 shows (out) two-tail probabilities for the Gaussian distribution. Hence, we can directly compare 2.5% (one-tail) and 5% (two-tail) probabilities in the respective tables. The Gaussian critical value for this probability is always $z^* = 1.96$, while the t-distribution critical value, t^*, depends on the number of degrees of freedom, v. The second column from the right in Table A-1 shows t^* as a function of v. When v is small, the difference between z^* and t^* is huge (e.g. $t^* = 12.7$ for $v = 1$). In such cases, the Student's t-distribution is significantly different from the Gaussian distribution. With increasing v, the critical t^* drops gradually, and eventually approximates the Gaussian critical value well. For a 'largish' $v = 30$, t^* is about 4% larger than the asymptotic Gaussian value. Your confidence intervals might be about 4% too small if you assume the Gaussian distribution for this sample size. With hundreds of degrees of freedom, the difference between the two distributions becomes negligible in practical applications.

Exercise 3.1

Presumably, the smallest division on the ruler is 1 mm, hence the reading error is ± 0.5 mm. However, when you measure the book, you actually take two measurements (i.e. at both ends) and find their difference. Most likely, one of them would be zero, because this is how you would place the ruler. But you still need to do two readouts, and each of them will contribute its own error. *Error propagation is explained in Chapter 7.* These errors add in quadrature, so the final measurement error is $\sqrt{0.5^2 + 0.5^2} \approx 0.7$ mm.

Modern books are cut to length with an accuracy far better than the ruler measurement error. Hence, if you were to measure many copies of the same book, you'd probably find exactly the same result each time. When subject-to-subject variability is less than the reading error, having multiple replicates does not make sense. However, it is very unlikely that you will encounter such a situation in biology, where things are seldom cut to length but rather show lots of variability.

Exercise 3.2

We can use square root of counts to estimate our uncertainties. These can be applied only to the original raw counts, but not to calculated rates. The raw murder numbers are $n_s = 6$ and $n_c = 19$, so their uncertainties are $\Delta n_s = \sqrt{6} \approx 2.4$ and $n_c = \sqrt{19} \approx 4.4$, respectively (here, s stands for Springfield and c stands for Capitol City). To find murder rates, each number is multiplied by a constant: a number of inhabitants divided by 100,000. When a *See Section 7.2 for error* number is scaled, its error scales in the same way, so fractional *propagation for scaling.* errors are conserved. The fractional errors for our cities are $\Delta n_s / n_s \approx 0.41$ and $\Delta n_c / n_c \approx 0.23$. Note that the error on the Springfield datum is rather large – just because the raw number is so small. When applied to murder rates, these fractional errors yield $\Delta r_s = r_s \Delta n_s / n_s \approx 1.7$ and $\Delta r_c = r_c \Delta n_c / n_c \approx 0.74$. Hence, the murder rates per 100,000 population with their errors are $r_s = 4.1 \pm 1.7$ and $r_c = 3.2 \pm 0.7$. These two errors overlap entirely ($r_c \pm \Delta r_c$ 'sits' inside $r_s \pm \Delta r_s$), and we can safely conclude that there is no statistically significant difference between them. A chi-square test confirms this with $p = 0.8$. The title of the 'murder capital' for Springfield is not justified.

Exercise 4.1

If we were to use the mean and standard error of the four results, we would find $D = (5.8 \pm 0.8) \times 10^{-3}$ μm^2 s^{-1}. However, this ignores individual errors and the fact that the first experiment collected more data and provided a more precise measurement. Instead, we should use the weighted mean, where weights are one over the error squared [equation (4-4)]. The weights for the four experiments are 2.8, 0.25, 0.04 and 0.25 (in units of 10^6 μm^{-4} s^2). Clearly, the first weight is an order of magnitude larger than the other weights, so the first experiment will dominate the combined result. Using equations (4-4) and (4-22), we can find the weighted mean and its error, $D = (4.5 \pm 0.5) \times 10^{-3} \ \mu m^2 \ s^{-1}$.

Exercise 4.2

Since $x_i = y_i$, we can replace all y_i with x_i in equation (4-15):

$$r = \frac{1}{n-1} \sum_{i=1}^{n} \left(\frac{x_i - M_x}{SD_x} \right) \left(\frac{x_i - M_x}{SD_x} \right) = \frac{1}{n-1} \sum_{i=1}^{n} \left(\frac{x_i - M_x}{SD_x} \right)^2$$

$$= \frac{1}{SD_x^2} \frac{1}{n-1} \sum_{i=1}^{n} (x_i - M_x)^2 = \frac{1}{SD_x^2} SD_x^2 = 1.$$

I used the definition of sample standard deviation [equation (4-8)].

Exercise 5.1

First, we need to calculate the mean, standard deviation and standard error for each sample. This is easy. Then, we find the critical value t^* for 95% confidence (2.5% one-tail probability), and $n - 1 = 11$ and 4 degrees of freedom for the control and treatment, respectively. Now, the confidence interval half-size is $t^* SE$, and the confidence interval for each sample extends from $-t^* SE$ to $+t^* SE$. All these calculation steps are shown in this table:

	Control	Treatment
M	1.030	0.571
SD	0.267	0.220
SE	0.077	0.099
t^*	2.201	2.776
$t^* SE$	0.170	0.274
95%CI	[0.86, 1.20]	[0.30, 0.85]

The 95% CIs for control and treatment do not overlap (just!). This suggests that perhaps there is a statistically significant difference between them. However, to assess this significance, you would have to do a proper statistical test (e.g. a t-test).

Exercise 5.2

Using the same approach as in the previous exercise, we find

	Day 1	Day 2	Day 1 + day 2
M	0.900	0.700	0.775
SD	0.0173	0.116	0.136
SE	0.0100	0.0518	0.0480
t^*	4.303	2.776	2.365
t^*SE	0.0430	0.144	0.114
95% CI	[0.86, 0.94]	[0.56, 0.84]	[0.66, 0.89]

Strangely, the 95% CIs from day 1 and day 2 do not overlap, so it is not easy to tell where the true mean is. Each CI gives us a 95% confidence that the true mean is within it. If it is in the day 1 interval, than we are unlucky to have the rare (one-in-20) day 2 experiment, where the true mean is outside the measured CI. And vice versa: if the true mean is within the day 2 interval, the day 1 experiment was unlucky. We could pool the data from two days together (they are, after all, replicates!) and get a better grip on the mean. The result is shown in the last column of the table above.

However, non-overlapping confidence intervals suggest a potentially important difference between the two days. A *t*-test provides a *p*-value of 0.03, which makes the difference rather significant (at least at the 'default' 0.05 level). One should exercise caution when interpreting these results, as it is quite likely that something had changed between day 1 and 2. Perhaps it was the temperature, humidity, instrument calibration or any other factor that might, or might not, have been under the experimenter's control. The changing factor introduced a systematic error, whereas the above calculations only find random errors. Sometimes even the most careful experimentalists make mistakes. It is not uncommon in biological experiments to see consistently different results, between two different days or two different cell cultures, or even between two different experimenters.

Exercise 5.3

Using equations from Sections 4.4 and 5.6, we find the following quantities (3 s.f.):

r	0.659
Z'	0.792
σ'	0.333
Z'_{lo}	0.138
Z'_{up}	1.445
r_{lo}	0.138
r_{up}	0.895
t	2.774
p	0.00983

The Pearson's correlation coefficient is high, but quite uncertain ($r = 0.7^{+0.2}_{-0.6}$, 95% CI). On the other hand, it is not consistent with zero, within error bars. This is confirmed by the high significance of correlation, $p = 0.01$. This all implies that the heights of fathers and sons are correlated. However, I would suggest a further study with a larger sample to confirm this finding.

Exercise 5.4

The proportions and the 95% CI calculated using the adjusted Wald method are as follows (data shown as percentages):

Strain	Proportion	Lower	Upper
1	11.6	9.8	13.8
2	6	1.5	17
3	40	12	78

They are all consistent with the true population proportion of 13.4%. It is worth pointing out that the second strain shows a rather low proportion of only 6%, but despite the relatively large sample size of 50, the errors are quite substantial. There is nothing particularly promising about this strain. Bad luck.

Exercise 5.5

We need to bootstrap the measurements and calculate the mean of each resampled sample. The distribution of these means approximates the sampling distribution of the mean. As I explained in Section 4.5, standard deviation of the sampling distribution of the mean equals the standard error of the mean. Hence, the standard deviation of bootstrapped means is an estimator of the standard error.

Exercise 5.6

Probabilities from non-overlapping intervals can be added together, so we can write

$$P(W < d) + P(d \leq W < n - d) = P(W < n - d).$$

This can be rearranged to find the required quantity:

$$\gamma_d = P(d \leq W < n - d) = P(W \leq n - d) - P(W < d).$$

The right-hand side of the equation can be calculated using an online probability distribution calculator. For the number of trials $n = 9$, the number of successes $d = 2$ and the probability of success in a single trial $p = 0.5$, we can find

$$\gamma_2 = P(W \leq 7) - P(W < 2) = 0.9805 - 0.0195 = 0.961.$$

Exercise 6.1

Figure S-1 shows three different graphical representations of the data. It is always good to plot the original measurements to see how they behave and spot any irregularities. In Figure S-1a, I have plotted T1, T2 and T3 versus WT. Note that due to the large span in values, I plotted logarithms of the data. Grey lines show the best-fitting linear relations. You can clearly see that there are two weird outliers in set T3, perhaps an indication of a problem with the experiment. You should probably go back to the lab and repeat the experiment, if possible. Beware: removing awkward

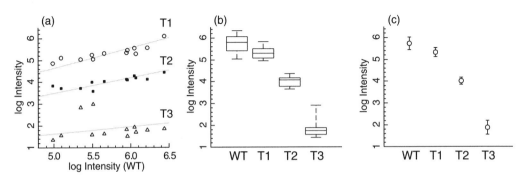

Figure S-1. Three different graphical representations of data from Exercise 6.1. (a) Raw intensities for each replicate plotted against wild type. Grey lines show best-fitting linear functions with no intercept. (b) Box plots. (c) Means with 95% confidence intervals.

measurements without making a clear statement in the publication is a serious breach of scientific rules and methodology. Simply speaking, it is cheating. You need to have a good reason to reject data, and it has to be explained in the paper. I therefore keep these two outliers for the purpose of this exercise.

Box plots are defined in Section 6.2.

Figure S-1b uses box plots. They show how the data are distributed. Again, I plot logarithms of the data. As the boxes are roughly symmetric, it suggests that the data are roughly log-normally distributed. The asymmetry of T3 is caused by the two outliers.

And, finally, Figure S-1c shows the mean and the 95% CI of the mean for each condition. These are useful for direct comparison between the conditions. For example, treatment T1 does not seem to be very successful, whereas T2 and, in particular, T3 reduce bacterial numbers very efficiently. This is only a graphical representation. In any case, an appropriate statistical test (e.g. a *t*-test) should be performed to assess the efficacy of each treatment.

Exercise 6.2

This is a tricky question. It is difficult to judge whether two means are significantly different just by looking at their standard deviations. This is why standard errors or confidence intervals are better for that purpose. Figure S-2 shows the original data, but with standard errors and 95% CI added. The *p*-values quoted in each panel come from the two-sample *t*-test, where the null hypothesis is that the two means are equal. A small *p*-value indicates significantly different means. Let us analyse this plot step by step.

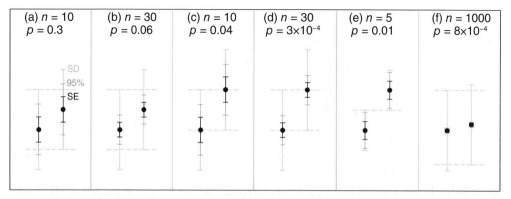

Figure S-2. This is Figure 6-12 with standard errors and 95% confidence intervals added. The statistical significance shown comes from a *t*-test.

Figure S-2a and S-2b look identical in the original figure. Their standard deviations are the same; they only differ in the sample size. From Figure S-2, we can see that standard errors overlap in S-2a but do not overlap in S-2b. 95% CIs do overlap in both cases. The means are not significantly different in either panel.

Figure S-2c and S-2d look the same in the original figure again. The figure above shows that their standard errors do not overlap at all. However, 95% CIs overlap in S-2c but do not overlap in S-2d. The means in S-2c do not seem to be significantly different, but they are different in S-2d with highly significant $p = 3 \times 10^{-4}$. Hence, 95% CIs seem to be a good indicator of different means.

Figure S-2e shows standard deviations that almost, but not quite, overlap, in contrast to Figure S-2d, where they overlap by about 50%. However, the means in S-2d are significantly different, whereas the significance is marginal in S-2e. This is because the sample in S-2e is very small. This demonstrates again that standard deviation is not a very good choice when you want to compare two or more samples.

Figure S-2f is a bit extreme. The standard deviations overlap a lot, but the sample size is large. The t-test shows that the two means are different with $p = 8 \times 10^{-4}$. Standard errors and 95% CIs are too small to be seen.

Exercise 7.1

The transformation function is $f(x_1, x_2) = x_1 x_2$. In order to apply the error propagation formula, we need to find derivatives:

$$\frac{\partial f}{\partial x_1} = x_2,$$

$$\frac{\partial f}{\partial x_2} = x_1.$$

From this and the propagation equation [equation (7-4)], we find

$$\Delta y^2 = x_2^2 \Delta x_1^2 + x_1^2 \Delta x_2^2 = x_1^2 x_2^2 \left(\frac{\Delta x_1^2}{x_1^2} + \frac{\Delta x_2^2}{x_2^2} \right)$$

$$= y^2 \left[\left(\frac{\Delta x_1}{x_1} \right)^2 + \left(\frac{\Delta x_2}{x_2} \right)^2 \right].$$

Here I assumed that $x_1 \neq 0$ and $x_2 \neq 0$. Dividing both sides by $y \neq 0$, we find

$$\left(\frac{\Delta y}{y}\right)^2 = \left(\frac{\Delta x_1}{x_1}\right)^2 + \left(\frac{\Delta x_2}{x_2}\right)^2,$$

which is identical to the ratio propagation formula (7-7).

Exercise 7.2

The radius, r, is measured with relative error of $\Delta r / r = 0.1$. The volume of the sphere is

$$V = \frac{4}{3}\pi r^3.$$

The derivative of the transformation function $f(r) = \frac{4}{3}\pi r^3$ with respect to r is

$$\frac{df}{dr} = 4\pi r^2.$$

Using the formula for error propagation (7-1), we find

$$\Delta V = 4\pi r^2 \Delta r = 3V \frac{\Delta r}{r}.$$

From which we can find the relative error:

$$\frac{\Delta V}{V} = 3\frac{\Delta r}{r}.$$

Hence, a 10% relative error in radius results in a 30% relative error in volume.

Exercise 7.3

The relation between NaCl mass, m; solution volume, V; molar concentration, c; and molar mass, M, is as follows:

$$c = \frac{m}{VM}. \tag{S-1}$$

The measurement errors are Δm and ΔV, and they propagate to the error of concentration using equation (7-7):

$$\left(\frac{\Delta c}{c}\right)^2 = \left(\frac{\Delta m}{m}\right)^2 + \left(\frac{\Delta V}{V}\right)^2. \tag{S-2}$$

We need to find m from the above equation. First, we find V from equation (S-1):

$$V = \frac{m}{cM}. \tag{S-3}$$

After substituting this into equation (S-2) and performing some simple algebra, we finally find

$$m = \frac{\sqrt{\Delta m^2 + \Delta V^2 c^2 M^2}}{\Delta c/c}. \tag{S-4}$$

The required molar concentration is $c = 0.01$ mol L^{-1} with accuracy $\Delta c = 10^{-4}$ mol L^{-1}. The molar mass of NaCl is $M = 58.443$ g mol^{-1}. Since the mass and volume can be read to the nearest milligram and millilitre, respectively, their errors are $\Delta m = 5 \times 10^{-4}$ g and $\Delta V = 5 \times 10^{-4}$ L (half of the smallest division). Substituting all this into equations (S-4) and (S-3), we find $m = 58$ mg and $V = 99$ mL. You can check this result by substituting these two numbers into equation (S-2). This yields the required $\Delta c/c = 0.01$. These are minimal values, so in a real experiment I would recommend that you at least double them, to make sure that the concentration accuracy requirement is met.

Exercise 8.1

The key to this problem is to find the relation between known mean, standard deviation and correlation coefficient, and the parameters described by equations (8-8). This is a very simple exercise, and by looking at definitions we find:

$$S_{xx} = (n-1)\,SD_x^2,$$
$$S_{yy} = (n-1)\,SD_y^2,$$
$$S_{xy} = (n-1)\,rSD_x SD_y.$$

I used the definition of Pearson's correlation coefficient [equation (4-15)]. Using the numbers from the exercise, we can find

$S_{xx} = 3.570$, $S_{yy} = 3.682$ and $S_{xy} = 3.163$. Then, using equations (8-7) and (8-12), we find the linear fit parameters and their standard errors: $a = 0.8860$, $b = -0.0228$, $SE_a = 0.1110$ and $SE_b = 0.05867$. The critical value from t-distribution with $n - 2 = 20$ degrees of freedom and $P = 0.025$ is $t^* = 2.086$ (see Table A-1). Hence, our best estimates for the slope and the intercept with 95% CIs are $a = 0.9 \pm 0.2$ and $b = 0.0 \pm 0.1$.

Exercise 8.2

There is a trap in this question. I asked for a 'linear regression', but the data are clearly not linear! See Figure S-3a. You can fit it with a straight line if you really want, but it is not going to make any sense. Data represent cell growth, so we rather expect an exponential dependence,

$$N = N_0 e^{\mu t},$$

where N_0 is a constant and μ is the growth rate. After taking a logarithm of this equation, we can establish a linear relation,

$$\ln N = \ln N_0 + \mu t.$$

The logarithm of cell count as a function of time is shown in Figure S-3b. It shows a clear linear relationship. Applying

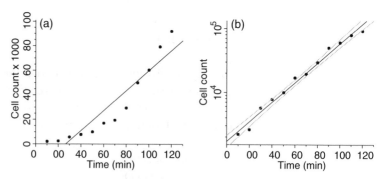

Figure S-3. Cell growth data in (a) linear and (b) logarithmic scale. Since we expect exponential growth of cell numbers, the logarithmic graph gives a linear relation between the two variables. Best-fitting straight lines are shown in black, and prediction 95% confidence intervals are shown in grey.

regression equations (8-7), (8-13) and (8-14) to $(t_i, \ln N_i)$ data yields

$$\mu = 0.035 \pm 0.003 \text{ min}^{-1},$$
$$\ln N_0 = 7.4 \pm 0.2,$$

where errors are 95% CI. The cell-doubling time, t_2, can be found by comparing the growth equation and $t = 0$ with $t = t_2$:

$$N_1 = N_0,$$
$$N_2 = N_0 e^{\mu t_2}.$$

From the doubling requirement $N_2 = 2N_1$, we find

$$t_2 = \frac{\ln 2}{\mu}.$$

By using this formula and propagating errors $(\Delta t_2 / t_2 = \Delta \mu / \mu)$, we can find

$$t_2 = 20 \pm 2 \text{ min}.$$

Appendix A

Table A-1. Critical values, t^*, for Student's t-distribution. For the given number of degrees of freedom, v (left), and for the given probability, p (top), the table shows the critical value t^* that cuts off the right-tail probability p from Student's t-distribution with v degrees of freedom: $p = P\,(T_v > t^*)$.

	Probability					
	0.0005	0.001	0.0025	0.005	0.025	0.05
1	636.6	318.3	127.3	63.66	12.71	6.314
2	31.60	22.33	14.09	9.925	4.303	2.920
3	12.92	10.22	7.453	5.841	3.182	2.353
4	8.610	7.173	5.598	4.604	2.776	2.132
5	6.869	5.893	4.773	4.032	2.571	2.015
6	5.959	5.208	4.317	3.707	2.447	1.943
7	5.408	4.785	4.029	3.499	2.365	1.895
8	5.041	4.501	3.833	3.355	2.306	1.860
9	4.781	4.297	3.690	3.250	2.262	1.833
10	4.587	4.144	3.581	3.169	2.228	1.812
11	4.437	4.025	3.497	3.106	2.201	1.796
12	4.318	3.930	3.428	3.055	2.179	1.782
13	4.221	3.852	3.372	3.012	2.160	1.771
14	4.140	3.787	3.326	2.977	2.145	1.761
15	4.073	3.733	3.286	2.947	2.131	1.753
16	4.015	3.686	3.252	2.921	2.120	1.746
17	3.965	3.646	3.222	2.898	2.110	1.740
18	3.922	3.610	3.197	2.878	2.101	1.734
19	3.883	3.579	3.174	2.861	2.093	1.729
20	3.850	3.552	3.153	2.845	2.086	1.725
25	3.725	3.450	3.078	2.787	2.060	1.708
30	3.646	3.385	3.030	2.750	2.042	1.697
40	3.551	3.307	2.971	2.704	2.021	1.684
50	3.496	3.261	2.937	2.678	2.009	1.676
100	3.390	3.174	2.871	2.626	1.984	1.660
1000	3.300	3.098	2.813	2.581	1.962	1.646

Degrees of freedom (left axis label)

Understanding Statistical Error: A Primer for Biologists, First Edition. Marek Gierliński.
© 2016 John Wiley & Sons, Ltd. Published 2016 by John Wiley & Sons, Ltd.

Table A-2. Confidence intervals for counts. There are three confidence intervals given in each row, 90%, 95% and 99% with 2 significant figures each. For example, a 95% CI for 5 counts is [1.6,12].

		90%		95%		99%	
		Low	High	Low	High	Low	High
	1	0.051	4.7	0.025	5.6	0.005	7.4
	2	0.36	6.3	0.24	7.2	0.1	9.3
	3	0.82	7.8	0.62	8.8	0.34	11
	4	1.4	9.2	1.1	10	0.67	13
	5	2	11	1.6	12	1.1	14
	6	2.6	12	2.2	13	1.5	16
	7	3.3	13	2.8	14	2	17
	8	4	14	3.5	16	2.6	19
	9	4.7	16	4.1	17	3.1	20
	10	5.4	17	4.8	18	3.7	21
	11	6.2	18	5.5	20	4.3	23
	12	6.9	19	6.2	21	4.9	24
	13	7.7	21	6.9	22	5.6	25
Count number	14	8.5	22	7.7	23	6.2	27
	15	9.2	23	8.4	25	6.9	28
	16	10	24	9.1	26	7.6	29
	17	11	25	9.9	27	8.3	31
	18	12	27	11	28	8.9	32
	19	12	28	11	30	9.6	33
	20	13	29	12	31	10	35
	21	14	30	13	32	11	36
	22	15	31	14	33	12	37
	23	16	33	15	35	13	38
	24	17	34	15	36	13	40
	25	17	35	16	37	14	41
	26	18	36	17	38	15	42
	27	19	37	18	39	15	43
	28	20	38	19	40	16	45
	29	21	40	19	42	17	46
	30	22	41	20	43	18	47
	40	30	52	29	54	26	59
	50	39	63	37	66	34	71
	60	48	74	46	77	42	83
	70	57	85	55	88	50	95
	80	66	96	63	100	59	110
	90	75	110	72	110	67	120
	100	84	120	81	120	76	130

Table A-3. Propagation of errors for a few commonly used transformations.

Function	Error
$y = ax$	$\Delta y = a\Delta x$
$y = ax^b$	$\dfrac{\Delta y}{y} = b\dfrac{\Delta x}{x}$
$y = a\log_b cx$	$\Delta y = \dfrac{a}{\ln b}\dfrac{\Delta x}{x}$
$y = ae^{bx}$	$\dfrac{\Delta y}{y} = b\Delta x$
$y = 10^{ax}$	$\dfrac{\Delta y}{y} = a\ln(10)\,\Delta x$
$y = ax_1 \pm bx_2$	$\Delta y = \sqrt{a^2\Delta x_1^2 + b^2\Delta x_2^2}$
$y = x_1 x_2, y = \dfrac{x_1}{x_2}$	$\dfrac{\Delta y}{y} = \sqrt{\left(\dfrac{\Delta x_1}{x_1}\right)^2 + \left(\dfrac{\Delta x_2}{x_2}\right)^2}$

Bibliography

Agresti, A., and B. Coull. 1998. Approximate is better than 'exact' for interval estimation of binomial proportions. *American Statistician*, 52, 119–26.

Brandt, S. 1999. *Data analysis: statistical and computational methods for scientists and engineers*. New York: Springer.

Cochran, W. G. 1977. *Sampling techniques*. New York: Wiley.

Cohen, J. 1988. *Statistical power analysis for the behavioral sciences*. Hillsdale, NJ: Lawrence Erlbaum.

Efron, B., and R. Tibshirani. 1993. *An introduction to the bootstrap*. New York: Chapman & Hall.

Fisher, R. A. 1970. *Statistical methods for research workers*. Edinburgh: Oliver & Boyd.

Gatz, D. F., and L. Smith. 1995. The standard error of a weighted mean concentration. 1. Bootstrapping vs other methods. *Atmospheric Environment*, 29, 1185–93.

Gehrels, N. 1986. Confidence-limits for small numbers of events in astrophysical data. *Astrophysical Journal*, 303, 336–46.

Gorard, D. 2005. Revisiting a 90-year-old debate: the advantages of the mean deviation. *British Journal of Educational Studies*, 53, 417–30.

Gurland, J., and R. C. Tripathi. 1971. Simple approximation for unbiased estimation of standard deviation. *American Statistician*, 30.

Hettmansperger, T. P., and S. J Sheather. 1986. Confidence-intervals based on interpolated order-statistics. *Statistics & Probability Letters*, 4, 75–79.

Hurlbert, S. H. 1984. Pseudoreplication and the design of ecological field experiments. *Ecological Monographs*, 54, 187–211.

Köbel, Jacob. 1535. *Geometrei. Von künstlichem Feldmessen und absehen*. Frankfurt.

Lem, S. 1979. *A perfect vacuum*. New York: Harcourt Brace Jovanovich.

Motulsky, H. 2010. *Intuitive biostatistics: a nonmathematical guide to statistical thinking*. New York: Oxford University Press.

Motulsky, H., and A. Christopoulos. 2004. *Fitting models to biological data using linear and nonlinear regression: a practical guide to curve fitting*. Oxford: Oxford University Press.

Olive, D. J. 2005. A simple confidence interval for the median. Accessed 8 January 2015. http://lagrange.math.siu.edu/Olive/ppmedci.pdf

Olkin, I., and J. W. Pratt. 1958. Unbiased estimation of certain correlation coefficients. *The Annals of Mathematical Statistics*, 29, 201–11.

Press, W. H. 2007. *Numerical recipes: the art of scientific computing.* Cambridge, UK: Cambridge University Press.

Sokal, R. R., and F. J. Rohlf. 1995. *Biometry: the principles and practice of statistics in biological research.* New York: W. H. Freeman.

Van Belle, G. 2008. *Statistical rules of thumb.* Hoboken, NJ: John Wiley & Sons.

Von Bortkiewicz, L. 1898. *Das Gesetz der kleinen Zahlen.* Leipzig: B. G. Teubner.

Index

actual response (in regression), 162

bar plots, 123–128
  with error bars, 126–128
bias, 52–53
binomial coefficient, 21
binomial distribution, 20–22, 36, 65, 87, 96, 184
bootstrapping, 103–105
box plots, 121–123

categorical variable, 9, 112, 121, 125, 130
central limit theorem, 16–18, 36, 68
centroid of the data, 164
Chauvenet's criterion, 16
chi-square distribution, 100, 178
chi-square statistic, 177
common uncertainty, 165, 173
confidence interval
  of the correlation coefficient, 90–95, 188
  of curve fitting parameters, 175–178
  of linear fit prediction, 170–172
  of the mean, 80–84
  the meaning of, 77–79
  of the median, 86–90, 190, 191
  of a proportion, 95–99, 184
  standard error as, 84–86
  why 95%?, 79–80
correlation coefficient, 63–64, 156, 188
  confidence interval of, 90–95, 188
  significance of, 93–95, 189
counting distribution, *see* Poisson distribution
counting error, 43–46
covariance, 149, 156, 170
critical *t* value, 82, 89, 168, 172, 185
cumulative distribution, 11, 27, 88
curve fitting, 175–178

degrees of freedom, 29, 59, 72–73, 81, 89, 94, 100, 164, 167, 172, 173
derivative, 144–148, 152–156, 163, 178
  partial, 146, 152, 156, 163
  and slope, 153, 154
dice, 8, 16

e-notation, 139
error bars, 112–141
  in bar plots, 126–128
  how to draw, 120–121
  overlapping, 128–130
  plots with no error bars, 130–132
error in the error, 71–72, 135, 136
error propagation
  correlated variables, 149
  difference, 146
  logarithm, 144–145
  multiple variables, 146–149
  product, 147–149
  ratio, 147–149, 186
  from replicated data, 150–151
  scaling, 144
  single variable, 143–145, 185
  sum, 146–147
errors
  asymmetric, 93, 98, 101, 119, 127, 185
  counting, 43–46
  measurement, 35–38
  random, 34, 50, 192
  reading, 41–43
  relative, 72, 135, 145, 148
  sampling, 39–41
  systematic, 33–34
estimator, *see* statistical estimator
expected value, *see* mean
explanatory variable, 113, 121, 131, 158, 175

factorial, 21
  of zero, 24
Fisher's transformation, 91
fitting
  curve, 175–178
  straight line, 162–164

Gaussian distribution, 13–16
  and central limit theorem, 16
gedankenexperiment, *see* thought experiment
geometric mean, 55–56
graphs, *see* plots

independent variables, 146, 149, 157, 170
interarrival times, 26–28

intercept, 158, 163, 171
    confidence interval of, 168
    standard error of, 166
intrinsic variability, 38–39

least-squares method, 163, 177
Levenberg–Marquardt algorithm, 177
linearization of data, 161
linear regression
    best-fitting parameters, 161–164
    errors of prediction, 170–172
    through the origin, 173–175
linear scale, 18, 117, 124
logarithmic plots, 117–118
logarithmic scale, 18, 117, 143
log-normal distribution, 18–20

margin of error, 97
mean
    geometric, 55–56
    of a random variable, 12
    of a sample, 53
    weighted, 54–55
mean absolute deviation, see mean
        deviation
mean deviation, 62–63
mean response (in regression), 159
measurement errors, 35–38
median
    in box plots, 122
    confidence interval of, 86–90, 190
    of a random variable, 12
    of a sample, 56–57

normal distribution, see Gaussian
        distribution

outlier, 15–16, 57, 122, 131, 190

Pearson's correlation coefficient, see
        correlation coefficient
pie charts
    don't make them, 128
plots
    bar, 123–128
    box, 121–123
    error bars in, 118–130
    labels, 116
    lines, 115
    logarithmic, 117–118
    symbols, 114
    with no error bars, 130–132
Poisson distribution, 23–28, 44,
        188
    interarrival times, 26–28
population, 12, 39, 47–49, 74
predicted response (in regression),
        162
probability distribution, 9–11
    binomial, see binomial distribution
    chi-square, see chi-square distribution
    continuous, 10–11

cumulative, 11, 27, 88
discrete, 9–10
Gaussian, see Gaussian distribution
log-normal, see log-normal
        distribution
Poisson, see Poisson distribution
t-distribution, see t-distribution
propagation of errors, see error
        propagation
proportion, 65, 184
    confidence interval of, 95–99,
        184
    standard error of, 97
pseudoreplication, 8, 106

radioactive decay, 23
random errors, 34–35, 50, 166, 192
    normal distribution of, 35–38
random events, 25, 28
random variable, 8–9
    continuous, 10
    discrete, 9
    mean of, 12
    median of, 12
    standard deviation of, 12
    variance of, 12
reading errors, 41–43
regression, 158
    linear, see linear regression
replicates, 74, 105–109, 191
    technical and biological, 192
resampling, 103
residuals, 12, 53, 162, 177
response variable, 113, 121, 158, 175

sample, 12, 39, 47–49, 74
sampling distribution, 75–77
    of the correlation coefficient, 91
    of the mean, 66
    of a slope in linear regression, 167
sampling error, 39–41
significant figures, 43, 84, 132–138
    definition, 132–133
    and errors, 135–136
    how to write, 133–135
SILAC (stable isotope labelling by
        amino acids in cell culture),
        187
slope, 153, 155, 158, 163, 169, 171
    confidence interval of, 167
    and derivative, 154
    standard error of, 166, 174
standard deviation
    of a random variable, 12
    of a sample, 57–59
    unbiased estimator, 59–62
standard error, 66–70
    as a confidence interval, 84–86
    of linear fit parameters, 166
    of a proportion, 97
    of the weighted mean, 71
standard score, see Z-score

statistical estimator, 49–52
  bias, 52–53
straight line fit, 161–164
Student's distribution, *see* *t*-distribution
systematic errors, 33–34

thought experiment, 44, 59, 66, 75, 81,
    166
true response (in regression), 160
t-distribution, 28–29, 69, 80–82, 89, 94,
    167–169, 172, 185
t-statistic, 29, 80, 94, 167

unbiased estimator, 52–53

variance, 12, 68, 73
  of a random variable, 12
  of a sample, 59–60
  unbiased estimator, 59–60

Wald method (adjusted), 97–99
weighted mean, 54
  standard error of, 71

Z-score, 13, 63